ROUTLEDGE LIBRARY EDITIONS: BRITISH SOCIOLOGICAL ASSOCIATION

Volume 21

PRIVATE RISKS AND PUBLIC DANGERS

PRIVATE RISKS AND PUBLIC DANGERS

Edited by
SUE SCOTT, GARETH WILLIAMS,
STEPHEN PLATT AND
HILARY THOMAS

Routledge
Taylor & Francis Group
LONDON AND NEW YORK

First published in 1992 by Avebury Ashgate Publishing Limited

This edition first published in 2018
by Routledge
2 Park Square, Milton Park, Abingdon, Oxon OX14 4RN

and by Routledge
711 Third Avenue, New York, NY 10017

Routledge is an imprint of the Taylor & Francis Group, an informa business

British Library Cataloguing in Publication Data
A catalogue record for this book is available from the British Library

ISBN: 978-1-138-49942-3 (Set)
ISBN: 978-1-351-01463-2 (Set) (ebk)
ISBN: 978-0-8153-4869-6 (Volume 21) (hbk)
ISBN: 978-1-351-16616-4 (Volume 21) (ebk)

Publisher's Note
The publisher has gone to great lengths to ensure the quality of this reprint but points out that some imperfections in the original copies may be apparent.

Private Risks and Public Dangers

Edited by

SUE SCOTT
GARETH WILLIAMS
STEPHEN PLATT
HILARY THOMAS

Explorations in Sociology No. 43

Avebury

Aldershot · Brookfield USA · Hong Kong · Singapore · Sydney

Published by
Avebury
Ashgate Publishing Limited
Gower House
Croft Road
Aldershot
Hants GU11 3HR
England

Ashgate Publishing Company
Old Post Road
Brookfield
Vermont 05036
USA

A CIP catalogue record for this book is available from the British Library and the US Library of Congress.

ISBN 1 85628 368 2

Details of other volumes in the *Explorations in Sociology* series are available from:

British Sociological Association Publications
Unit 3G, Mountjoy Research Centre
Stockton Road
Durham DH1 3SW

Printed and Bound in Great Britain by
Athenaeum Press Ltd., Newcastle upon Tyne.

Contents

Acknowledgements

Many people have supported the production of this volume, but our main thanks go to two people: Anne Dix, the British Sociological Association's Administrative Secretary whose knowledge, organizational skill and support made organizing the BSA's 1991 conference a much more pleasant experience than it would otherwise have been; and Nick Dalziel, the BSA's Publications Officer, for the hard work he has put into negotiating the publication of this volume.

Foreword

Ronald Frankenberg
Brunel and Keele Universities

> And instead of being angry, I began to giggle at his absurdity, at the X-ray sister's cruelty, and at my own helplessness. The three conditions moved together to make a fantastic picture in my mind. It was as if some gigantic buffoonery had descended upon the whole earth.
>
> I laughed for all the horrors I could imagine existing at that moment; for all the terrors and despairs and tortures and madnesses. I laughed to think of the strange things that were happening to human bodies all over the world. Some were feasting, some were locked in dungeon cells, some were copulating, some were singing, some were having babies, some were starving, some were weeping, some were having Turkish baths, and some were being torn to bits. I thought of the infinite number of postures, expressions, gestures and functions of the body; and these were the funniest thoughts of all. (Denton Welch 1985 (1950) *A Voice Through a Cloud*, London: Penguin Books p. 52)

The general metaphor for standing aloof, watching from a distance, is significantly *clinical* detachment. The clinic is, as Foucault (1973) and Armstrong (1983) have taught us, the place increasingly remote from lived life, where biological life is maintained by experts who impose their will upon our bodies and our minds. But is it detached from danger, remote from risk. It is our cultural duty to be grateful for, and our social duty to resent, their attentions. We recognize their distinctions as simultaneously salutary and dangerous, especially those between what can be observed through enhanced sight and what can be felt through enhanced sensitivity of feeling and emotions.

In the past, Uta Gerhardt and myself have severally preached that the point of intersection of the sociology of health and illness with general sociology was the contempt of experts for the mass which imperialism and Fabianism had synergistically engendered (*sic*). Significantly on this grander, dare one say post-modern, occasion, we both started from

Auschwitz. She told us that progress deserted culture there; I suggested that on the contrary, following Bauman (1989) and the film *Shoah*, that Auschwitz was the *reductio ad absurdum* of enlightenment modernist doctrine (Spiegelman 1987 and 1992). The Holocaust and the murderous abuses of Stalinism were triumphs of *disembodied* rationalism. Jewish and Kulak, gypsy and gay, lame and handicapped are characteristics seen as undesirable and to be removed in as efficient a way as possible; the ultimate solution is the ultimate technological fix. Some even describe them metaphorically as cancerous on the body social or the body politic. The fact that they are characteristics of living people, bodies personal, who are simultaneously destroyed is less than relevant to such thinking.

Romanyshyn (1989) contrasts for us a view of Florence before the invention of perspective, with one painted shortly afterwards. In the first, we feel that we might well bump into the walls or into fellow human beings. We can almost smell what Italians call in contrast to the odour of sanctity, *humanitas*. In the second, we are distanced. The buildings are seen in orderly ranks and too far away to know if they are even inhabited; the first stage of a process that was to lead to aerial photography and the Registrar-General's classifications. He puts these in apposition with two portraits of anatomists alongside the human body. In the first, the scientist embraces and looks at the body whole and little detail is seen within it; in the second, Vesalius is depicted displaying the interior of one dismembered arm and stands beside it, holding but not looking at it. He looks at us, Little Jack Horner who has just discovered the plum. He has destroyed the perduring living body as the basis of life and medicine and invented the corpse, as well as the mere somatic fragment. He has laid an uncertain but enduring foundation for the lay/professional contradiction within health care (including that between nurse and doctor – 'nurses did not give post-mortem care to dead bodies. Instead they cared for patients they knew both before and after death', Wolf 1988 p.68).

In the new general atheory of total relativism ($E = WMC^2$) which follows the supposed death of metanarrative Marxism and Freudianism, the absence of universal consensus with white middle class and Western male culture (WMC^2), is interpreted as a total emptiness (E) of theory rather than a need to pose enduring questions in a new way. From the dust created by this explosion of overdetermined structural theory the living world of humanity has to be re-created.

For as Bryan Turner (1984, 1987) predicated and as this collection shows, medical sociology has embodied, in its practices (with their focus on detail) and embedded the questions it poses (with their focus on specificity), problems which are central to the whole sociological enterprise. The subject matter of the suffering, living mindful body, its ills and its healing, its triumphs and its defeats is a major challenge to the mere statistical aggregation of attributes, whether by epidemiology or by the sociological grand surveys of the 'abstracted empiricists' (Wright Mills 1959). To make its relevance clear to our colleagues in other fields of sociology, however, we do need theory, perhaps not in the form of metanarratives but certainly in the form of metaquestionnaires. We cannot always provide the right answers but we do need to aspire, at once more modestly and more arrogantly, to pose the right questions. We can seek them perhaps in the neglected work of Roslyn Bologh (1979, 1981), in that of Robert Crawford (1984) and in the well know but incompletely assimilated work of Anselm Strauss (1985).

Bologh (1981) characterizes biomedical relationships as marked by the alienation of patients as a class from medical knowledge, by their powerlessness and by the corresponding separation of self and body. She suggests to us the questions: 'Who owns medical knowledge and what is it about?', 'In what sense is the implied division of property: the physician has the mind, the nurse has the body, the patient the duty to obey, the clue to healing?' Crawford writes on the neglected topic of the culture of health rather than of illth, i.e. sickness, illness, disease, affliction or suffering. Strauss and his colleagues chart and analyze for us all the many kinds of work from sentimental to pharmacological and technological, which affect the trajectory of illness. The implications of their analyses of this work done, by patients and their relatives in hospital (household and clinic) with a little help from their professional friends, is for us to seek the common thread in almost all these tasks: In what ways and by whom is body work practised?

Lifedeath (Frankenberg 1987), as is implied in the epigraph from Denton Welch, is the chronic sickness *par excellence* and our colleagues need to recognize as we do that it is dominated much of the time by body work. Not merely, of course, work on the body but also work by the body. The body is not merely good to think with, as Mary Douglas suggests, or a symbolic field to mirror or reproduce dominant values and conceptions; it is also a site for resistance to, and transformations of, imposed meaning. As Crawford (1984) argues cogently and empirically, cultural meanings are not only shared and given, they are fragmented and contested. Social life is divisive as well as cohesive. The body makes, and is made by, a fractured social world (Bourdieu 1977; Csordas 1990; Merleau-Ponty 1989).

The title of the volume, *Private Risks and Public Dangers*, and the editors' introduction draw attention to the current version of mind/body dualism which encourages the individual to see her/his body in separation from self and in isolation from society. Risks are something an individual takes; dangers are something an individual avoids. This opens the possibility, even the compulsion, of viewing damage to the body corporeal, as merely the concern of its separately perceived owner. This is an underlying theme of all the chapters, clearly illustrated by those on Foetal Alcohol Syndrome, the sexual behaviour of gay men and HIV testing technology. Until challenged, as it is by Judy Green and by Helen Roberts and Charlie

Davison and their respective colleagues, such dualism makes plausible the Rational Choice theory view that health is achieved merely by communication from us, the mindful knowledgeable, to the hiterto ignorant but wilful them. To continue to suffer is in these terms merely to continue to be wilful rather than mindful. It is not just the specialist sociology of health and illness that needs to recognize that the incarnate body (in contrast to the merely biological) is the living site of a contested hegemony of power. The analyses below forcefully reveal this. Sociology at large can learn from them, perhaps, that the expertise of sociologists therefore lies in its own destruction by fostering recognition of the university of embodied expertise.

References

Armstrong, D. (1983), *Political Anatomy of the Body*, Cambridge: Cambridge University Press.

Bauman, Z. (1989), *Modernity and the Holocaust*, Ithaca, NY: Cornell University Press.

Bologh, R. (1979), *Dialectical Phenomenology: Marx's Method*, London: Routledge & Kegan Paul.

Bologh, R. (1981), 'Grounding the alienation of self and body: A critical phenomenological analysis of the patient in Western medicine', *Sociology of Health and Illness*, **3** (2), 188–206.

Bourdieu, P. (1977), *Outline of a Theory of Practice*, Cambridge: Cambridge University Press.

Crawford, R. (1984), 'A cultural account of "health": control, release, and the social body', in McKinlay, J.B. (ed.), *Issues in the Political Economy of Health Care*, New York and London: Tavistock.

Csordas, T. (1990), 'Embodiment as a paradigm for anthropology', *Ethos*, **18**: 5–47.

Foucault, M. (1973), *The Birth of the Clinic*, New York: Viking.

Frankenberg, R. (1987), 'Life: cycle, trajectory or pilgrimage? A social production approach to Marxism, metaphor and mortality', in Bryman, A., Bytheway, B., Allat, P. and Keil, T. (eds), *Rethinking the Life Cycle*, London: Macmillan.

Merleau-Ponty, M. (1989), *Phenomenology of Perception*, London: Routledge.

Romanyshyn, R.D. (1989), *Technology as Symptom and Dream*, London and New York: Routledge.

Spiegelman, A. (1987, 1992), *Maus: A Survivor's Tale, Parts 1 & 2*, London: Penguin Books.

Strauss, A. Fagerhaugh, S., Suczek, B. and Wiener, C. (1985), *Social Organization of Medical Work*, Chicago: University of Chicago Press.

Turner, B.S. (1984), *The Body and Society: Explorations in Social Theory*, Oxford: Blackwell.

Turner, B.S. (1987), *Medical Power and Social Knowledge*, London: Sage Publications.

Wolf, Z.R. (1988), *Nurses' Work, the Sacred and the Profane*, Philadelphia: University of Pennsylvania Press.
Wright Mills, C. (1959), *The Sociological Imagination*, Harmondsworth: Penguin Books.

1 Introduction

Sue Scott and Gareth Williams

> Keep your hands off! Go to your Bacchic rites, and don't
> Wipe off your crazy folly on me. But I will punish
> This man who has been your instructor in lunacy.
> (Euripides, *The Bacchae*)

The chapters in this volume were selected from a large number of interesting papers presented at the British Sociological Association's annual conference in March 1991. The theme of the conference was 'Health and Society' and it was organized by the editors on behalf of the BSA. In constructing the conference 'streams', and in inviting Uta Gerhardt and Ronnie Frankenberg to be our plenary speakers, we were concerned that the conference should illuminate the connections between the sociology of health and illness and the central concerns of sociology (Turner 1987). During the planning of the conference we heard rumours to the effect that it would be of interest only to medical sociologists; as if by being 'about' health, the conference could not also be 'about' society. In the event these suppositions proved ill-judged. The conference attracted sociologists from many fields along with participants from other disciplines such as politics, anthropology, social policy, geography, epidemiology, and a significant number of health care workers.

All the contributions collected here underline the importance of strengthening the links between sociology of health and illness and both general sociology and related disciplines. In examining aspects of 'risk' and 'danger' in both private life and the public domain, the contributions to this volume deal with interdisciplinary issues at the interface between individual behaviour and social structure. Neither risk nor danger featured in the titles of the conference streams, but on reading the papers in order to make our selection, we discovered that these issues cut across a number of the streams, and connected with a range of current debates in sociology.

A risky society

In many ways our society has become a much less secure place than it was twenty years ago. This is not the place to engage in party political debate, but the editorial columns of the *British Medical Journal*, for example, wish to leave no doubt about where the causes lie:

> Legionnaire's disease, salmonella, AIDS, drug misuse, homelessness, pollution and environmental blight are all rampant – symptoms of the deprivation and despair created by divisive socioeconomic policies. Once more the inner city is crumbling into a health and welfare nightmare, redolent of Dickensian horrors (Porter and Porter 1990)

Whatever one's view of the politics underlying this observation, it does serve to draw our attention to the social context of health hazards and the inequalities in exposure to them. Moreover, in referring us back to the nineteenth century, it reminds us of the close relationships between public health, civil society and the State, particularly in cities.

It is in the urban setting, above all, that the respectable classes are exhorted to be alert to those Dionysian excesses which apparently endanger biology, morality and society all at the same time. The Manchester conference was held only a mile or so down the road from the location of the nineteenth century slum housing so pungently described by Engels in *The Condition of the Working Class in England*, and the picture of ill-health and social deprivation in inner Manchester in the late twentieth century remains grim (Manchester City Council 1987, Manchester Family Health Services Authority 1990). The conference site was also only a couple of miles from an urban zone which has recently been labelled 'Britain's Bronx'. While the reportage undoubtedly contains some hyperbole (Taylor 1991), Manchester has apparently become a dangerous place, a city of drugs and guns where the population is constantly at risk (Cohen 1991).

The urban environment as a place of danger and hazard is considered very differently in the opening and closing chapters of this volume. Gavin Kendall and Gary Wickham explore the emergence of the discourse within which ideas about the city as a threat to health are located, while Helen Roberts and her colleagues examine accidents to children and the strategies developed by parents to maintain child safety in an urban setting fraught with danger. From very different angles both chapters confront the role of expert knowledge in the 'government of populations', and the way in which communities or members of 'social bodies' can resist or oppose experts' answers to their 'problems' by asking different questions framed in their own terms. What is of central concern to sociologists, in contrast perhaps to their colleagues in public health, is not the evaluation of risk and danger in any absolute sense, but rather our shifting perceptions of risk and changing patterns of risk management (Douglas 1985).

Risk and rationality

It has been suggested that '... to live in the universe of high modernity is to live in an environment of chance and risk' (Giddens 1991, 109). This may, at first glance, seem to be both blindingly obvious and somewhat ahistorical. Life is and always has been a risky business, but with longer life spans (in the West at least) people would seem to be less prone to the vicissitudes of a risky environment than ever before. However, we inhabit a culture highly sensitive to risk and, as Armstrong (1991) has pointed out, the threats come from everywhere – from the air that we breathe, the rays of the sun, the multi-national petrochemical companies, the 'man' in the street, from our families, our sexual partners. ... even the cells of our bodies may turn against us.

These anxieties must be understood, as Judy Green shows in her chapter on accidents, in the context of a rational discourse which established the expectation that both nature and people could be controlled for the greater good, either through science, or surveillance, or both. The concept of risk we now use, based on individualism rather than fate, is relatively modern; and we look in the main to expert systems and the State to keep danger at bay, rather than surrendering ourselves to the Will of God or to Nature. Since the mid-1970s, however, there has been an ideological shift from the rights of individuals to have their health protected by the State, to the responsibilities of individuals to protect themselves from risk. In the managed society, danger is (in theory) controlled for us, but as good citizens we must play our part by managing our own relationship to risks through self-surveillance (Department of the Environment 1990) .

The chapters by Kendall and Wickham, Green and Richard Freeman, in his critical review of the 'idea of prevention' in health and social policy, all make an important contribution to understanding how it is that certain issues come to be defined as manifestations of danger, risk, or hazard. These forms of knowledge, we are now accustomed to hearing, are about power; not just in the domination of some groups by others, but in the way in which the institutional order is created and sustained. The idea of prevention can almost be understood as a metaphor for attempts to manage risk in relation to health, and the history of prevention can teach us a great deal about prevailing ideologies of health and the changing relationship between the individual and the State.

Private risks and public dangers

Since the development of public health as a specialist field in the latter part of the nineteenth century, the gaze of the experts has moved from the environment (water, sewers etc.) through the spaces between people (infection and contagion) to the effects of behaviours on individual bodies (Armstrong 1991). The construction of 'knowledges' or discourses serves to 'problematize' both public issues and, as Vikki Bell shows in her chapter, certain kinds of intimate relations such as incest. We have seen an increasing focus on the family both as a site of dangers and as an institution at risk. Bell's chapter, and that by David Clark and David Morgan on marriage guidance, describe discourses in which various experts are invoked to comment upon the dangers lurking within what is supposed to be the safe, trustworthy environment of family and home. Luhman has suggested that 'trust is the solution to a specific problem of risk' (1986, 95), and the modern period has seen the development of a discourse of trust in the context of intimate relationships alongside the existing (if dwindling) trust in experts and expert systems. However, if trust is the basis of intimate relationships, then it is very difficult constantly to call it into question (Holland et al 1991). The papers by Clark and Morgan and Graham Hart and his colleagues indicate some of the issues which arise when the gaze of the experts focuses on the sphere of the intimate (Weeks 1991).

The relationship between lay and expert knowledge is a concern of a number of the contributions. The chapter by Charlie Davison and his colleagues, for example, explores the interface between the professional directives to look after ourselves and the cultural context within which individuals make sense of both the 'risks' to their health, and the exhortations to change their behaviour. The chapter by Hart and his colleagues scrutinizes the concept of 'relapse', as it is applied to the seeming inability of some gay men to maintain the 'safer sex' strategies which they had developed in the context of the threat of HIV infection. The authors clearly indicate that decisions about risk-taking are linked to the context of people's everyday lives and not to some set of absolute health messages. Both papers offer clear illustrations of the need for sociological analysis in areas usually dominated by epidemiology.

Individualistic rational choice theories, beloved of many health policy makers, lead us away from the complexities and contradictions of both our everyday lives and the wider culture into a punitive, victim-blaming environment. The papers by Maureen McNeil and Jacquelyn Litt, and Hart et al, illustrate this focus on recalcitrant individuals, and depict how this process divorces behaviour from its social context, and treats risk-taking as wilful deviance rather than as the product of a complex blend of identity, intersubjectivity, and structural location.

One way of managing anxiety about risks is to see them as something which affects other people and to construct real or symbolic *cordons sanitaires*. The focus on so-called risk groups in the context of HIV infection is a clear illustration of this. Evan Willis's paper elucidates some of the issues involved in such attempts to extend the surveillance apparatus in order to create, for some, an aura of safety. Through an analysis of the development of a moral panic about Foetal Alcohol Syndrome, McNeil and Litt demonstrate how a particular group, in this case native American women, come to be defined as a danger to society. The authors reveal the ways in which the medical gaze has extended its focus to the point where it 'penetrates' the womb. As in the chapter by Bell, this case-study describes in close detail the ethical and political issues that emerge from the exposure of personal lives to public scrutiny through expert discourse.

Many of these chapters deal with what one might call internal loci of control, and the ways in which societal agencies attempt to inculcate self-restraint in order to reduce risky behaviour. The contribution from Joel Richman and Tom Mason on the future of the Special Hospitals, deals with a 'risk group' who are regarded as so dangerous, so 'out of control', that the only acceptable mode of restraint is external, visible, and often highly repressive (Brindle 1992). By way of an examination of the organizational setting and professional discourses surrounding 'psychopathy', this chapter asks us to think about what is to be done with individuals who are cut off from society and held in the 'bastion of society's nightmares' (Richman and Mason, Chapter 10) in order to protect others from the risk of harm (Foucault 1988).

A sociology of risk

In this book we have brought together papers on diverse aspects of risk, danger, and their regulation in an attempt both to display their variety and to draw attention to the common threads which link them. Many of these contributions, in one way or another, concern those classical Dionysian indulgences – food, drink, and sex – which, in certain situations at certain times, are seen to threaten public health and public order. They relate directly, we would argue, to questions of 'governmentality' (Foucault 1991, Rose 1990), and it may be in relation to the health implications of self-pleasuring that 'private life', as we understand it in Western societies, is likely to be extinguished in the name of public safety. In their inability or unwillingness to acknowledge the social origins and social construction of public health risks, agents of social control, to use an old-fashioned expression, blame and restrain individuals in a variety of ingenious ways. The response of the State, and some sections of civil society, to the risks and dangers described in many of these chapters is, like the Apollonian reaction to the Dionysiac cult in *The Bacchae*, fearful and punitive: an epidemic psychology (Strong 1990).

It is the role of the sociologist to render problematic the constructions of social problems which are presented by policy makers, the media and other opinion formers; and to explore both the historical contexts in which such issues evolve, and the part they play in people's everyday lives. The chapters collected here accomplish this in a variety of ways. We hope they will make a contribution to a more sympathetic and sociologically informed debate about the private risks which individuals take and the inequalities and differences shaping the public dangers they are expected to manage.

References

Armstrong, D. (1991), 'Spaces, boundaries, and power in public health', paper presented at the British Sociological Association, Medical Sociology Conference, University of York, September 1991.

Brindle, D. (1992), 'Mental patients terrified by pig's head at "repressive hospital"', *The Guardian*, 10 March.

Cohen, N. (1991), 'Grim routine of violence in the poverty trap', *The Independent*, 29 April.

Department of the Environment (1990), *This Common Inheritance*, Government White Paper, London: HMSO.

Douglas, M. (1985), *Risk Acceptability According to the Social Sciences*, London: Routledge.

Foucault, M. (1988), 'The dangerous individual', in Kitzman, M. (ed.), *Michel Foucault: Politics, Philosophy Culture*, London: Routledge.

Foucault, M. (1991), 'Governmentality', in Burchall, G., Millar, P. and Gordon, C. (eds), *The Foucault Effect*, Brighton: Harvester.

Giddens, A. (1991), *Modernity and Self Identity*, Oxford: Polity.

Holland, J., Ramazanoglu, C., Scott, S., Sharpe, S. and Thomson, R. (1991), 'Risk, power and the possibility of pleasure: young women and safer sex', *AIDS Care*, Spring.

Luhman, N. (1986), *Love as Passion*, Oxford: Polity.

Manchester City Council (1987), *Poverty in Manchester: the Third Investigation*, Manchester City Council.

Manchester Family Health Services Authority (1990), *Health Profile of Manchester*, Manchester: FHSA.

Porter, D. and Porter, R. (1990), 'The ghost of Edwin Chadwick', *British Medical Journal*, **301**: 252.

Rose, N. (1990), *Governing the Soul: the Shaping of the Private Self*, London: Routledge.

Strong, P. (1990), 'Epidemic psychology: a model', *Sociology of Health and Illness*, **12**: 249–59.

Taylor, I. (1991), 'Moral panics, crime and urban policy in Manchester', *Sociology Review*, **1**: 28–30.

Turner, B.S. (1987), *Medical Power and Social Knowledge*, London: Sage.
Weeks, J. (1991), 'Rethinking private life', Jackie Burgoyne Memorial Lecture, Sheffield City Polytechnic, December.

2 Health and the social body

Gavin Kendall and Gary Wickham

Introduction

In considering the *health* of *social bodies*, we are primarily concerned with the role of knowledges in the modern era (late eighteenth century onwards) in the government of populations. By 'populations' we mean groupings of individuals and organizations bounded in some way, usually by geography, but also by such factors as age, sex, mental health, and religion.

The government of populations involves what Foucault calls 'governmentality'. He defines this in the following terms:-

> the ensemble formed by the institutions, procedures, analyses and reflections, the calculations and tactics that allow the exercise of this very specific albeit complex form of power, which has as its target population, as its principal form of knowledge political economy and as its essential technical means apparatuses of security (Foucault 1979:20).

More specifically, governmentality involves processes by which particular populations are first defined and then subjected to definite political strategies. These strategies are potentially infinite, as indeed are the targeted populations. In addition to strategies which aim at the medical health of populations, other governmental strategies aim at making target populations more economically healthy (Miller and Rose 1990; Rose 1990), more mentally healthy (Miller and Rose 1986), or more educationally healthy (Hunter 1990).

There is no inherent maximum or minimum size of target populations and no limit to the scope of governmental strategies – 'government' can be about families being governed to be more internally self-regulating in terms of their health (which, remember, can be physical, mental, economic)

(Donzelot 1979), congregations being governed to be more spiritually healthy, as well as nation states or even 'the world' being governed to be more militarily healthy, or more environmentally healthy. What is important for such a theory of 'government' is that the populations under consideration are conceptualized in terms of definite conditions of their operation – they have definite means of being calculated – while the corresponding political strategies are also conceptualized in terms of definite conditions of operation – they too have definite means of being calculated, expressed and carried out (by definite agents or forces).

In all of this, knowledges are crucial. As we indicated earlier, our focus is on the role of knowledges in the processes of government. Specific knowledges make possible both the calculation and the targeting of different populations. For example, in the field of education, knowledges of education, civilization and literature, among many others, are combined in the formulation and execution of strategies to make the educated population 'more healthy' by making them particular sorts of citizens (Hunter 1989, 1990; Rose 1984), while knowledges of education, economics and politics, among many others, are combined in the formulation and execution of a strategy to make target populations (teachers, children, students, lecturers) 'more efficient' as a way of making them 'more healthy'.

This essay has three main sections. In the first, we elaborate our understanding of 'the social body' and use this understanding to offer some remarks about cities as social bodies. In the second, we elaborate our understanding of 'health' and its coarticulation with social bodies. In the third, we apply our descriptions of health and social bodies to cities.

Our essay is meant as an indication of the usefulness of Foucault's notion of governmentality for expanding studies of 'health' and 'cities'. In this sense, we are proposing a framework for describing the health of cities which respects the heterogeneity of 'health' and 'cities'.

Social bodies and cities

In trying to come to terms with the role of knowledges of social bodies in government, we have in mind Turner's history of the metaphor of the body: 'The body was ... used as a ... general metaphor for the structure and function of society as a whole' (Turner 1988:177). We are not concerned here with an exact genealogy of such uses, rather we follow Turner only to garner a working definition of social bodies.

Building on this, we, like Turner, turn to Foucault, who addresses the invention of 'society as a whole':

> What was discovered at ... the end of eighteenth century ... was the idea of *society*. That is to say, that government not only has to deal with a territory,

with a domain, and with its subjects, but that it also has to deal with a complex and independent reality that has its own laws and mechanisms of disturbance. This new reality is society ... It becomes necessary to reflect upon it, upon its specific characteristics, its constants and its variables ... (Foucault 1989:261, emphasis in original).

In other words, what was developed here was a particular knowledge device to aid the government of more and more populations, particularly larger populations. The development of 'society' as such a device was occasioned by a shift in ways in which the body could be conceptualized:

> [I]t's the body of society which becomes the new principle in the nineteenth century. It is this social body which needs to be protected, in a quasi-medical sense. In place of the rituals that served to restore the corporal integrity of the monarch, remedies and therapeutic devices are employed such as the segregation of the sick, the monitoring of contagions, the exclusion of delinquents (Foucault 1980:55).

The idea of 'the social body', then, was (and is) very useful for governing this new object – 'society'. In governing society, another knowledge device paralleling and variously connected to 'the social body' was the idea of 'the city':

> One begins to see a form of political literature that addresses what the order of a society should be, what a city should be, given the requirements of the maintenance of order; given that one should avoid revolts, permit a decent and moral family life, and so on. In terms of these objectives, how is one to conceive of both the organization of a city and the construction of a collective infrastructure? ... There is an entire series of utopias or projects for governing territory that developed on the premise that a state is like a large city; the capital is like its main square; the roads are like its streets ... The model of the city became the matrix for the regulations that apply to a whole state (Foucault 1989:258–60).

Cities, therefore, as social bodies, were and are objects of knowledge, used to govern or manage particular populations – whether entire state populations, entire urban populations, or populations as small as congregations, classrooms or families – in order to make them 'healthier' in some way or other.

Health for social bodies

As we suggested in the introduction, by 'health' we mean much more than a medical condition of particular persons. What interests us are the ways in which certain knowledges, which we group together under the term 'health', are used in the assessment of different populations for different governmental strategies. That is to say, we are proposing a framework for the description of those processes of government which allow more and more groupings of people and organizations – populations – to be judged healthy or unhealthy (or somewhere in between) by teachers, social workers, public and private bureaucrats, parents, economic advisers, military planners, as well as by doctors. These actors make their judgements in terms of the strategies they oversee – for example, parents judge their children healthy or unhealthy in terms of their adherence to a set of familial norms; military planners judge their existing and potential armed personnel healthy or unhealthy in terms of their capacity to wage offensive or defensive wars. Other terms may be used as substitutes/synonyms for the continuum which runs from 'healthy' to 'unhealthy' – 'normal' to 'abnormal' or 'pathological'; 'efficient' to 'inefficient'; 'hygienic' to 'unhygienic'; 'prepared' to 'unprepared'; 'well behaved' to 'disturbed'; 'proper' to 'improper'.

We do not mean to suggest that doctors are unimportant in judging the health of particular populations but the medical establishment is only one of many sets of institutional and technological resources involved in governing health, in the broad sense of the term we are arguing for here. Doctors are indeed technicians of the social body, as we intimate below, but they are only some of many such technicians. Particularly in the nineteenth century, these technicians began to understand the problem of health through its relation to two sorts of dangerous phenomena: the dangerous classes and the dangerous individual.

The nineteenth century saw the gradual but thorough elaboration of a notion of class, especially in terms of topographical and historical patterns of analysis (Jones and Williamson 1979). What is important for our purposes is the notion of a class as an entity with its own inherent qualities to do with health – in the case of the 'dangerous classes', an inherent quality as dangerous to their own health, and, more particularly, to the health of others.

'Dangerousness' (in our analysis, danger to 'health') is closely related to the possibility of crime, which in turn is seen to flow from the unhealthy conditions in which such classes existed:

> In Manchester and the manufacturing districts, as everywhere else, the class of criminals spring chiefly from the most abject, ignorant, improvident and poverty stricken of the population; and the quarters which they occupy are the lowest, the most incommodious and loathesome in the city ... Here, then, the narrow, ill-ventilated, and filthy streets, with their stunted and filthy hovels, markedly contrast with the spacious warehouse, the lofty factory, and the public buildings situated in the more airy and commodious quarters of the town; and while such is the uninviting character of the external appearance, it is only exceeded by the confinement, darkness, nakedness, and filth which characterise the thickly populated dwellings of a class who live by a mingled recourse to elemosynary

relief and criminal pursuits, and whose honest or illicit gains are spent with equal improvidence and profligacy ... It is from such a region of physical degradation that contagion and pestilence spread abroad, and it is out of such a community that crime has its birth, and principally emanates (William Beaver Neale, *Juvenile Delinquency in Manchester: its causes and history, its consequences, and some suggestions concerning its cure*, as cited in Jones and Williamson 1979:83, originally published in 1840).

This notion of dangerous classes as a particular danger to health was articulated in relation to those classes who were 'at risk' from them: the 'perishing classes'. A mutual relation was thus established between two classes which were viewed as having intrinsic qualities or characteristics, as either dangerous or at risk in terms of their health. The theoretical establishment of such classes enabled the elaboration of notions of treatment and prevention, frequently understood in terms of public hygiene. One obvious example of such 'treatment' or 'prevention' was the establishment of compulsory education, which sought to make appropriate interventions that were guarantees of health.

Such strategies at the level of class also had analogies at the individual level: the 'dangerous individual'. This figure, according to Foucault, was constituted at the intersection of law and psychiatry (Foucault 1988). 'Dangerousness' came to inhere in the characters of the criminals themselves, rather than the crime committed; the correlate of the dangerous individual, the 'at risk' individual, can now become visible. These concepts aim at the maintenance of public health by the calculation of individual danger and risk, yet we can see that this health is guaranteed through a variety of disciplinary interventions: psychiatry and law are the most obvious here, but it should be clear that medical, topographical, economic, and political knowledges are also important.

The notion of health with which we are working may be connected to one of two understandings of health in the Hippocratic writings (produced roughly between 430 and 330 BC). The *Tradition in Medicine* suggests that the body has a natural state of health which is disturbed by illness. Health, then, is understood teleologically, as the doctor attempts to move the body towards a state of perfect health. Such a reading of health is particularly noticeable in the Aristotelian and Galenic traditions, which are also of crucial importance in the formulation of modern ethics of health.

The other understanding of health contained in the Hippocratic writings can be seen in *Epidemics III*. This takes a much less teleological approach to its subject matter; some diseases are regarded as untreatable, and part of a natural life cycle. Health is less an ultimate state characterized by the absence of illness, than a set of various conditions of the body: even death can be health. Such a non-teleological understanding of health can be seen in some modern versions of homeopathy, but this resource is almost invisible in the modern age.

The first of the above two Hippocratic understandings, as it is applied to the health of the individual body, has contributed to the metaphorical content for the expanded notion of health. It is in this sense that this particular understanding is currently dominant in knowledges of health. For such knowledges, the health of a city, for example, is measured against some imaginary ideal situation, a point at which a city would be in perfect health. This utopianism, however, does not allow the possibility for any other disposition of the city to be regarded as 'healthy'. For example, a city would not be allowed to 'die'.

The health of cities as social bodies

> In the eighteenth century, the development of demography, of urban structures ... had raised ... the question of human 'populations', with their conditions of existence, of habitation, of nutrition, with their birth and mortality rate, with their pathological phenomena ... The social 'body' ceased to be a simple juridico-political metaphor (like the one in the *Leviathan*) and became a biological reality and a field for medical intervention. The doctor must therefore be the technician of this social body, and medicine a public hygiene (Foucault 1988: 134).

The growth of the city was a condition of possibility for the type of health strategies we are interested in here. It was only in the eighteenth century that births began regularly to exceed deaths, so it is not surprising that it was necessary to wait this long for urbanization on a similar scale to that of antiquity (Braudel 1981:72, 483ff). In the pre-modern era, by contrast, there were very few cities of any great size in the West: Paris had a population of 300,000 in 1565 (Benedict 1989:69), while Venice had a population of around 170,000 in 1563 (Chambers 1970:123).

For us, urban growth in the modern era is most importantly an expansion in knowledges to do with urban 'health'. The eighteenth century in England

> saw the acceleration of the modern torrent of writing, discussion, and analysis, as well as prints, drawings, and designs, relating to the understanding and improvement of the urban environment (Corfield 1982:168).

This period also saw the beginning of official interest in the longevity and mortality rates of the urban populations in Europe (Lampard 1976; Hacking 1990), although the first official census was not taken in England until 1801, and the boom in official statistics did not begin until the latter part of the nineteenth century (Braudel 1981:64ff).

Architecture, as a set of knowledges and as a practice, was given a considerable boost in eighteenth century England by the abundance of capital and low rates of interest. Building and rebuilding of the urban environment were characteristic of this period (Corfield 1982:168), a process which was to gather pace in the nineteenth century, but primarily in the private rather than the public sphere. The number of architects in London doubled in twenty years early in the century because of a world-wide trend in which investors began to see great financial benefits in providing rental slum dwellings for the urban poor (Hobsbawm 1977:212).

The division of the city into areas dealt with by police and health boards began to allow for the possibility of seeing a link between squalor and crime, such that a strong link between a kind of physical health and a kind of moral hygiene began to be enunciated. The science of sanitary engineering was rapidly becoming of crucial importance, but its 'hygienic' (health) value was of a double character, equating the provision of physical hygiene with a more thoroughgoing 'health' in the population. The supply of water on a regular basis to the large cities was an enormous problem, but one which in its solution 'enabl[ed] cities of unprecedented size and complexity to develop'. As such, '[t]he real heroes of the age were not the general run of doctors, politicians or businessmen, but a forgotten group of sanitary reformers' (Cannon 1975:164).

The urban areas of nineteenth century Britain were still rather dangerous places to live, with frequent outbreaks of diseases which killed thousands. Much energy was expended on sewage and water-supply systems, as mentioned above, but there were also attempts to remove insidious pollutants. For example, the Health of Towns Association was formed in 1844 to address the problem of factory smoke (Briggs 1968:375). Another, less obvious, pollutant derived from the decaying dead. As Cannon notes, in Australia, some cemeteries had to be re-sited:

> as late as the 1880s, Sydney people were still burying their dead in the Devonshire Street cemetery, "in the midst of a bustling community", where in rainy weather "slimy and offensive matter" oozed through the stone walls into Elizabeth Street (Cannon 1975:145ff).

The growth of the science of statistics in the nineteenth century enabled the police and health boards to notice patterns of disease and crime, and to conclude that these two must somehow be linked, assuming a topographical link to be evidence of a causal link. This is perhaps the beginning of an understanding of the relation between the individual and the state which emerged by the end of the nineteenth century and which sees an intimate link between the political health of the nation and the 'health' of the individual, considered in the broad terms we have been using throughout. This is a refinement of an earlier view which, from at least the seventeenth century, had linked the political health of the nation with the condition of the inhabitants of that nation (Tribe 1978). This refinement, this 'individualization', of the members of the population, seems to be something which relates to a new way of calculating individuals and populations, particularly as put into practice in the human sciences which emerged from the middle-to-late nineteenth century (Rose 1985).

Assessments of the health of cities were sometimes deeply affected by technical accidents. The invention of the elevator, at the end of the nineteenth century, for example, was a technical breakthrough which, quite unknown to its inventors, was a condition of possibility for a new conception of the urban. The elevator, *inter alia*, made possible the proliferation of tall buildings in north American cities, and the gradual emergence of new city shapes. Despite, or perhaps because of, the accidental nature of such factors, the twentieth century saw the explosion of a new formal knowledge or science which aimed at 'capturing' or 'taming' the whole of cities in order to make them healthy in the broadest sense, by urban planning. The twentieth century has experienced a 'glut of city planners' (Briggs 1968:381). With a few notable exceptions (Paris, Washington, Stockholm and a few others), it was rare before our century began for great cities to be planned. However, the twentieth century has seen many new cities developed from a plan and many old cities made subject to a plan – Chicago (1909), Moscow (1935), Amsterdam (1934), New York (1926), Rome (1931, 1940–1), Sydney (1945), Toronto (1949), Manchester (1945) and London (1943) (Robson 1954:88ff; Briggs 1968:381ff).

The city, as we have hinted, is a site of concern for moral health. Throughout the eighteenth and nineteenth centuries in England, cities were feared as places of disorder, due to contagion. The design and the policing of urban areas were subjected to scrutiny in terms of ensuring that disorderly gatherings – known loosely as 'riots' – which constituted legitimate expressions of 'public opinion', did not develop to the stage of 'threats' (Corfield 1982:161).

These possible threats are also dealt with in the general context of public health. For example, in 1765, Manchester inaugurated a system of Police Commissioners for 'cleansing and improving the streets' (Corfield 1982: 158). In doing this, Manchester was continuing a mediaeval obsession with keeping streets 'clear of encumbrances' and 'clear of dirt'. In Paris, similar tactics involved 'street widening, street straightening, demolition of balconies, projecting oriels, and even occasionally whole buildings' (Girouard 1985:65, see especially chs 12, 14). But it is clear that these hygienic tactics also operate in relation to other aspects of the health of populations. We notice a gradual tactic emerging in the nineteenth century of attempting to break up areas of known criminal activity by routing new roads through them, as in the example of the St Giles area of London, which saw the routing of New Oxford St through it (Tobias 1967). Similarly, the building of Central Station in Manchester was partly aimed at a known rookery.

Measures designed to have beneficial effects on the health and hygiene of populations may have unintended consequences. The provision and regulation of utilities in England in the nineteenth century, such as paving, cleansing, and the supply of water, also crucially included street lighting. This led to an 'artificial' extension of the hours of public activity, and the gradual development of the notion of urban 'night life' (Corfield 1982: 168–9; Cannon 1975:116 ff). Street lighting can be seen in relation to the insertion of pleasure into city life, such that city life is no longer simply constituted around the working day.

To take another example: it has been argued that modern city welfare administration measures have regulated cities to such an extent that they have directly led to population increases and overpopulation, and hence to the expansion of capitalist industrial enterprise, and hence, by implication, to larger cities with greater need for welfare administration, and so on *ad infinitum* (Jacoby 1973:61). Thus this aspect of the urban is a self-generating circle – whether vicious or virtuous is a matter for debate.

Conclusion

We have highlighted the importance of particular knowledges of health for governing urban life. Both contingency and technical innovation are crucial in this. One finds a picture of contingent connections between different knowledges and urban health rather than a picture of grand and overarching plans for domination of the population. These connections in turn make new ways of government possible and make visible new problems to be addressed by these new ways of government.

We are very wary of any suggestion that urban environments are governed, in relation to 'health' or anything else, in any completely successful way. The various obstacles to the maintenance of the health of cities as social bodies may be overwhelming. For example, the regulation of cities has had to overcome 'famines, plagues, civil wars, and the complete demoralization of economic life' (Abbot and Johnson 1926:192; Braudel 1981:74). Many critics are even sounding the death knell for the city: Mumford, for instance, argues that modern cities are 'economically unsound, politically unstable, biologically degenerate, and socially unsatisfying' (Robson 1954:102), while Newman (1986) argues that Liverpool is likely to become the first modern city to be abandoned.

We are not suggesting that the health of the social body is ungovernable, that it cannot be regulated at all. We merely wish to oppose the idea that health and the social body are or can be regulated in a complete, successful sense.

We thank our colleagues on the larger project of which this essay is part – Graham Crist, Hannah Lewi, and Bill Taylor – for their help in preparing the essay. We also thank Doug Fraser, Mike Michael and Martin O'Brien for their comments on an earlier draft. We acknowledge the funding assistance of Murdoch University and the University of Western Australia.

References

Abbot, F. and Johnson, A. (1926), *Municipal Administration in the Roman Empire*, Princeton: Princeton University Press.
Benedict, P. (1989), *Cities and Social Change in Early Modern France*, London: Unwin Hyman.
Braudel, F. (1981), *The Structures of Everyday Life: the Limits of the Possible*, London: Collins.
Briggs, A. (1968), *Victorian Cities*, Harmondsworth: Pelican.
Cannon, M. (1975), *Life in the Cities*, Melbourne: Thomas Nelson.
Chambers, D. S. (1970), *The Imperial Age of Venice 1380–1580*, London: Thames and Hudson.
Corfield, P. J. (1982), *The Impact of English Towns 1700–1800*, Oxford: Oxford University Press.
Donzelot, J. (1979), *The Policing of Families*, New York: Pantheon.
Foucault, M. (1979), 'Governmentality', *I&C*, **6**: 5–21.
Foucault, M. (1980), 'Body/power', in Gordon, C. (ed.), *Michel Foucault: Power-Knowledge: Selected Interviews and Other Writings 1972–1977*, New York: Pantheon.
Foucault, M. (1988), 'The dangerous individual', in Kritzman, L. (ed.), *Michel Foucault: Politics, Philosophy, Culture: Interviews and Other Writings 1977–1984*, New York: Routledge.
Foucault, M. (1989), 'An ethics of pleasure', in Lotringer, S. (ed.), *Foucault Live: Interviews 1966–84*, New York: Semiotext(e).
Girouard, M. (1985), *Cities and People: a Social and Architectural History*, New Haven: Yale University Press.
Hacking, I. (1990), *The Taming of Chance*, Cambridge: Cambridge University Press.
Hippocrates, *The Tradition in Medicine*.
Hippocrates, *Epidemics*.
Hobsbawm, E. (1977), *The Age of Capital 1848–1875*, London: Abacus.
Hunter, I. (1989), *Culture and Government: The Emergence of Literary Education*, Basingstoke: Macmillan.
Hunter, I. (1990), 'Personality as a vocation: the political rationality of the humanities', *Economy and Society*, **19** (4): 391–430.
Jacoby, H. (1973), *The Bureaucratization of the World*, Berkeley: University of California Press.
Jones, K. and Williamson, K. (1979), 'The birth of the schoolroom', *I&C*, **6**: 59–110.
Lampard, E. (1976), 'The urbanizing world', in Dyos, H. J. and Wolff, M. (eds), *The Victorian City: Images and Reality, Volume One: Past and Present/Numbers of People*, London: Routledge and Kegan Paul.
Miller, P. and Rose, N. (eds) (1986), *The Power of Psychiatry*, Cambridge: Polity.
Miller, P. and Rose, N. (1990), 'Governing economic life', *Economy and Society*, **19**, (1): 1–31.

Newman, P. (1986), 'Lessons from Liverpool', *Planning and Administration*, 1986–7.
Robson, W. (1954), *Great Cities of the World: their Government, Politics and Planning*, London: Allen and Unwin.
Rose, J. (1984), *Peter Pan: or, the Impossibility of Children's Fiction*, Basingstoke: Macmillan.
Rose, N. (1985), *The Psychological Complex: Psychology, Politics and Society in England, 1869–1939*, London: RKP.
Rose, N. (1990), 'Governing the enterprising self', in Heelas, P. and Morris, P. (eds), *The Values of the Enterprise Culture – The Moral Debate*, London: Unwin Hyman.
Tobias, J. J. (1967), *Crime and Industrial Society in the Nineteenth Century*, London: Penguin.
Tribe, K. (1978), *Land, Labour and Economic Discourse*, London: Routledge and Kegan Paul.
Turner, B. (1988), *The Body and Society*, Oxford: Blackwell.

3 Some problems in the development of a sociology of accidents

Judy Green

What is an accident?

Milk is spilt, a car crashes, a woman unintentionally becomes pregnant. Accidents will happen, and presumably the misfortunes (and, occasionally, happy events) which we label accidents always have happened. Accidents form a heterogeneous and loosely defined category of events. They are identified not by their end state (such as a death or an injury) but by their ascribed cause, or rather, lack of cause. We do not decide that an accident has occurred by observing what happened, but by investigating how it happened.

If we want to describe the rules for including an event as an accident there are two features that are apparently characteristic, and these will form an initial working definition. The first is that an accident should be an unmotivated event. Neither the victim nor any other agency, human or divine, willed it to happen. Modern ironic usage clearly illustrates this: 'We could arrange for you to have an accident' (Fo 1980). In general no-one can be blamed for an accident. It is this feature which distinguishes accidents from wilful damage and neglect. The arrangement of physical objects and temporal sequences that precede an accident must be seen to be purely coincidental: they cannot have been willed. The causation of accidents is arbitrary and not logical.

Second, and following from this, an accident is unpredictable as a unique event. Although the epidemiology of accidents can be mapped, the occurrence of a particular accident cannot be foreseen. The victim, in an 'ideal' accident, has no previous knowledge of the misfortune and therefore cannot be held responsible. From an early age we learn to negotiate claims to the accidental in order to claim or disclaim responsibility. Anyone with children will recognize the claim for clemency: 'I didn't do it on purpose – it was an accident'.

Accidents and rationality

I would like to argue here that accidents have not always happened; that they are the product of a modern and Western rational discourse and that they constitute a marginal category of that rational explanatory system. In everyday discourse, accidents are the events that are left over when no motivation, culpability, or predictability can be attributed after an event has been subjected to moral scrutiny. As a category of misfortune, accidents have been somewhat neglected as a legitimate object of theoretical concern. The aim of this paper is to suggest some of the problems in the development of such a theory. Medicine and the social sciences have been concerned with accidents, but often only in so far as they can be 'rationalized': that is, that they can be brought within the same explanatory framework as other events, with identifiable causes and a regular pattern of occurrence. An adequate sociological theory of accidents would also account for their essential features in the lay classification system: namely unpredictability and lack of motivation.

Research on accidents has fallen into two broad categories. Both deny one of these essential features. The first is the epidemiological approach which seeks, often explicitly[1], to render the accidental predictable and, by implication, avoidable. Statistics on accidents in the home, on the road or in the work place can be examined to reveal broad patterns and underlying causes. With this magnification, the notion of the accident as unique event is rendered illusory; as random noise within a larger regular pattern, and the victim is constructed as possibly culpable for not preventing it. More recently, studies of accidents in industry have criticized the 'victim-blaming' of much health and safety research (Tombs 1991) but have still seen accidents as largely inevitable and predictable (Perrow 1984).

A second kind of approach is at the other end of the spectrum in terms of scale and seeks to provide meaning for individual accidents in the symbolic relationship they may have to events and relationships in the victim's life. The work of Freud (1938) on the psychology of the accident, in which accidents are seen to be the epiphenomenal indicators of deep structures, is illustrative. Again common sense notions of the accidental are not explored but rather dismissed as illusory. Accidents are conceptualized here to be specifically the result of motivation, if unconscious. I will return to the work of Freud, but note here that these approaches, however fruitful at one level, seem inadequate as starting points for a sociology of the accident as part of a classification of misfortune, which takes account of the place such events have in the ways we make sense of the world. It may be more useful to attempt first to identify when events such as those described above became classified as 'accidental'.

The emergence of the accident

The modern concept of the accidental first appears in England at the end of the seventeenth century. Hacking (1975) has admirably discussed the emergence of the science of statistics in the West at this time, arising from changing discourses of probability and evidence, and although he is not looking at accidents specifically, his ideas provide a clue to the emergence of a modern notion of accidental happenings. By the end of the seventeenth century, Hacking argues, it became possible to talk in terms of risks and probability as old ideas of external evidence (the authority of learned authors) which became replaced by internal evidence, whereby deductions could be made from the evidence of things and the relations between them. 'Statistics', claims Hacking, 'began as the systematic study of quantitative facts about the state' (1975:102). By the second half of the seventeenth century, social records such as death rates could be examined as evidence for truths, rather than merely as 'signs' of some event such as a plague epidemic. Hacking compares the different kinds of evidence provided about the plague by Thomas Lodge (writing in the beginning of the century), a writer and poet who went on to study medicine, with that of John Graunt and William Petty, two of the first statisticians to examine death rates, later in that century. Whereas Lodge writes of the symptoms of the plague and warnings of its arrival as mere 'signs' and as traces of the plague whose validity is backed up by testimony 'gathered out of the most approved authors' (Lodge 1603:A4), Graunt and Petty can examine the bills of mortality for internal evidence, and can make inferences based on probabilistic logic from the data itself. I am going to use examples from these three writers and one other, Samuel Ward, a controversial preacher of Ipswich who died in 1640, to illustrate how this new discourse of probability also produced the accident as a marginal category of misfortune.

To start with the word 'accident' itself, the OED gives one meaning that is now obsolete; that of an 'occurring symptom'. Thomas Lodge clearly uses the word in this sense:

> The most troublesome or dangerous accident in this Sicknesse are weakness of vertue, faintings of the heart, soundings, raving or frenzie ... (Lodge 1603:K2)

There is no moral content to the term here: accidents are merely the deteriorations or 'events' in the course of the illness and nothing more.

So what were happenings with no logical or willed cause labelled? It appears that there was simply no need to distinguish them from those more culpable acts or those with some definable motivation. Samuel Ward, for instance, in a sermon entitled *Woe to Drunkards*, lists the many misfortunes that befall those who drink. These are attributed to the direct judgement of God, to whom drunkenness is an odious sin, punished by all manner of untimely ends. He lists numerous examples of such misfortunes, including deaths from diseases we would describe as arising from alcohol use, such as a man who

> having surcharged his stomacke with drinke, hee fell vommiting, broke a Veyne, lay two days in extreme paine of body and distresse of mind, till in the end recovering a little comfort, he died. (Ward 1622:24–25)

Other misfortunes include injuries sustained while drunk, including a

> man 85 yeares old, or thereabout, in Suffolk, overtaken with Wine [who] going downe a paire of staires ... fell, and was so dangerously hurt, as hee dyed soone after, not being able to speake from the time of his fall to the time of his death. (Ward 1622:29)

Although drunkenness is seen to be an apparent cause of this and other fatalities, it is not the ultimate cause, which is God's retribution. Misfortunes which are not associated with the victim's own drunkenness are therefore also subsumed under the same explanatory framework: thus a woman who has persuaded three men to stay to drink some more is 'suddenly taken speechlesse and sicke, her tongue swolne in her mouth, never recovered speech, the third day after dyed' (Ward 1622:20).

We would now perhaps divide such misfortunes into three distinct categories with very different moral content: effects of disease as a result of alcohol use, injuries arising from drunkenness (both of which are seen as somewhat blameworthy) and true 'accidents'; injuries sustained by people who may be drinkers but who were not inebriated at the time. Those injuries not attributable to the causal effects of alcohol use we would have no ready explanation for, but within the moral universe of Ward's God there is no need for such a category since the judgement of God is an all embracing explanation. The sin of drunkenness is the ultimate cause of all the misfortunes to which drinkers are prone, through the intervention of a righteous God. There is no need, with such a universal explanatory framework for a category of events for which no cause is known.

By the 1660s though, a dramatic change has occurred. We find the beginnings of the concept of a category of events which are not regular and which do not fit within a predictable pattern. With the Renaissance, rational ways of thinking were beginning to gain precedence, and the study of disease and mortality was no exception. Graunt, writing in the second half of the seventeenth century, has an approach to investigating causality that we would recognize as essentially 'rational':

> ... among the several casualties some bear a constant proportion unto the whole number of Burials; such are the chronic diseases, and the diseases whereunto the City is most subject; as for example Consumption, Dropsies, Jaundice,

Gowt, Stone, Palsie, Scurvy, rising of the Lights ... rickets, aged, agues, Feavers, Blood Flux, and Scowring: nay, from Accidents, as Grief, Drowning, Mens' making away themselves and being killed by several accidents &c do the like, whereas Epidemical and Malignant diseases, as the Plagues, Purples, spotted Feaver, Small Pox and Measles do not keep that equality, so in some Years or Months there died ten times as many as in others (Graunt 1662:18).

So here, although the term 'accident' still refers to a group of happenings distinguished by the suddenness of their effect or the externality of the cause of death rather than their exclusion from a pattern, there is the notion of happenings which do not 'bear a constant proportion to the whole'. They are not yet called accidents, but a recognizable notion equivalent to our one of the accidental is clearly present. The events which do not fit in Graunt's scheme are the epidemics, for which he had not established a predictable pattern from his study of the Bills of Mortality. Graunt later lists some of the events which we might now label as accidents, claiming that it is not worth examining them too closely as they are not amenable to the discovery of regular patterns:

We shall say nothing of the numbers of these that have been Drowned. Killed by falls from Scaffolds, or by carts running over them &C, because the same depends on careful Trade, and Employment of men and upon matters which are but circumstantial to the Seasons, and Regions we live in; and affords little of that Science and Certainty we aim at. (Graunt 1662:23)

Graunt's project in these *Observations* quoted above was in some senses emblematic of the beginnings of the process of rationality. It was an attempt to calculate the risk of various kinds of death: a project continued with great enthusiasm today (British Medical Association 1990). He appeals to empirical evidence to calculate the exact chance of death from various causes:

... whereas many persons live in great fear, and apprehension of some of the more formidable and notorious diseases following; I shall only set down how many died of each: that the respective numbers, being compared to the total 229250, those persons may the better understand the Hazard they are in ... (Graunt 1662:16)

By the end of the seventeenth century Petty, also writing on mortality statistics, could confidently dismiss arbitrary or non-rational causes as having no relevance to his scientific project. Reviewing changes in the population of London, he writes:

what reason to assign the like increase from 1604 to 1642 I know not, unless I should pick out some Remarkable Accident happening in each part of the said

> Period and make that to be the Cause of this Increase (as vulgar People make the Cause of every Man's Sickness to be what he did last eat) (Petty 1699:27)

'Remarkable Accidents' are here causes not worthy of serious consideration. Using the new found rationality in political statistics, Petty could now separate what had for Graunt been irregular features, such as plague years, and place them neatly within the larger pattern of predictable population shifts. He predicts, for instance, the time it will take for London's population to double by calculating population increases based on bills of mortality and

> including some allowance for Wars, Plagues and Famine, the effects whereof, though they may be Terrible at the Times and Places where they happen, yet in a period of 360 years is no great matter in the whole Nation (Petty 1699:18)

At the beginning of the century there was no need to appeal to the accidental; by the end it was not only a necessary but also a despised mode of explanation.

Medical classifications

The medical classification of accidents has implicitly accepted these lay definitions of accidents that emerged at the end of the seventeenth century. The moral content of the categorization was finally made formal by the statistician William Farr in the Registrar General's 16th Report of Births, Marriages and Deaths, published in 1856. He introduces a new classification for violent deaths:

> Human agency plays so important a part in this class, that it might be made into the basis for orders. Thus a man may die 1. a glorious death in battle (pro patria mori); he may die 2. by an act of homicide (murder, manslaughter); he may die 3. ignominiously on the scaffold (execution); or 4. abandoning the post where God has placed him, he may take away his own life (suicide); 5. he may die by a surgical operation 6. he may die by accident. If this grouping be adopted, the mode in which the death is produced by wounds, chemical injuries, poisons, asphyxias, and mechanical forces, would form secondary heads. (Registrar-General 1862:78)

The importance of these distinctions lies in the purely moral categorization of the meaning of the death: whether glorious or ignominious; whether the victim or another was culpable. Those with no discernible moral content are the 'left-overs', the accidents. The medical accident was now defined to mirror lay terms. A fatal accident was an event for which there was apparently no motivation, but which lay on a boundary between the need for a cause (in which case the coroner's court would be asked to attempt to provide one) and the lack of a 'real' cause as defined by the infant science of statistics.

The modern International Classification of Disease Categories (WHO 1977) maintains the spirit of Farr's original distinctions. Accidents are in the last category, Order XVII, which, unlike the other classes, is not based on a system of the body or a group of disease causing agents. Reporting on deaths in this class the Registrar General divides Order XVII, Deaths from Violence, along moral lines into deaths caused by others (homicide) deaths caused by the victim (suicide) and deaths from which no fault can be attributed: the accidents. Although since the ninth revision of the International Classification of Disease Categories (WHO 1977), the main subdivision of accidental death is ordered by medical outcome (a fractured skull, for example). A subsidiary classification based on the social and environmental context of the accident persists, allowing deaths from accidents to be classified by, for instance, where they occurred, or how they happened. This is implicit recognition that an accident is defined by more than its medical sequelae: that information is needed also about *how* it was caused.

Medicine, it seems, treats accidents with a degree of embarrassment. They are relegated to the end of a classificatory system; treated in casualty departments which are often badly resourced and which have an ambiguous status within the hospital (Calnan 1982) and their victims and witnesses are often cross examined with minute detail to determine whether they really are entitled to be treated as if there was no moral culpability involved. This process of moral interrogation is at its most public in the coroner's court, which provides the official categorization of a death as accidental (Green 1992).

The attribution of moral content relies on the 'objective' evidence of physical traces, as well as verbal reconstruction of events leading up to the accident. It is in forensic science the study of revealing culpability achieves its most sophisticated in terms of practical application. Post-mortem study of wounds can reveal not only the medical cause of death, but also where the blame should lie: as a text book of forensic medicine puts it, unlike the wounds of homicide and suicide 'which follow certain traditional rules ... accident is something unforeseen. It is not planned and does not therefore develop along orthodox lines' (Simpson and Knight 1985:68). The accident is what is left unaccounted for after any of the regular wounds of motivated death are found.

Social accounts of the accidental

If nosology has relegated the accidental to the end of the classification system, sociology and other social sciences have done no better. As suggested earlier, epidemiological studies have attempted to impose a pattern from above, ignoring the social meaning accidents have as unique events. Psychology has inferred a pattern on an individual level in uncovering the workings of the unconscious mind. In *The Psychopathology of Everyday Life*, Freud (1938) argued that the minor mishaps that we label meaningless and accidental are signs of the ordered rational workings of the unconscious: they are only superficially 'accidental', with the real meaning lying beneath the surface to be revealed by the analyst. Mishaps, 'slips of the tongue' and even trivial losses have meaning:

> Whoever forgets articles in the doctor's office, such as eye glasses, gloves, handbags, generally indicates that he cannot tear himself away and is anxious to return soon (Freud 1938:155f).

In this case it is that the patient wishes to remain with their analyst. The losses of personal possessions dear to us are not trivial accidents but are the effects of the unconscious working out perhaps difficult relationships or are representations of the manipulation of meaningful symbols. This attribution of rationality to the seemingly irrational is, for Freud, a comfort:

> It is consoling to think that the 'losing of objects' by people is merely the unsuspecting extension of a symptomatic action, and is thus welcome at least to the secret intention of the loser (Freud 1938:154).

That the inexplicable has been brought within the realm of the explicable is an advance in knowledge: for a rationalist, there can be no mere coincidences left to clutter a universal explanatory framework.

'Slips of the tongue' likewise reveal more meaning than the patternings of superficially correct speech. Although this explanation of 'Freudian slips' has entered lay theories of accident causation, it seems doubtful whether it has changed the definitions that operate to include or exclude events from the category of accidents. It has merely excluded a certain group of happenings from the category in that we may cease to see as accidents what there is now a rational causal explanation for, even if that explanation relies on appeal to unconscious motivation. As the Freudian unconscious has become part of lay theories of causation it has become possible to see motivation as being hidden from the actor. Denial of motivation is no longer enough to make a successful claim for a speech accident to have happened (indeed the very denial might furnish definitive proof of unconscious motivation). Denial merely places the speech accident on the negotiable boundary space of morally loaded and motivated actions. Freud's work may have shifted the boundaries of the accidental (using the name of one's previous lover to the new one may no longer be forgivable as a purely accidental utterance) but it does not seem to have dispersed the category of events that are deemed accidents.

In terms of examining the social meaning of accidents the major work has again come from the field of psychology. Rotter's influential development of the internal-external locus of control construct examined the extent to which ideas of fate and chance (external factors) were important to individuals, rather than 'internal factors': beliefs in self determination and the ability to control one's own destiny (Rotter 1982:145). Although Rotter claimed that the external pole of his continuum was not by definition the negative one (Rotter 1982:272) and that scores on either end of the continuum indicate pathological beliefs, the implication of his writing is that the external pole is the negative one. The definition of an external locus of control is a belief in the power of 'luck, fate, chance or powerful others' to control the outcome of behaviour. In his original article Rotter quotes Veblen on a belief in luck being characteristic of 'primitive' societies and claims that behaviour based on 'accidental' reinforcements will lead the subject to 'learn the wrong things and develop a pattern of behaviour which Skinner has referred to as 'superstitious' (Rotter 1966)

In the light of such clear evaluations it is difficult to reconcile Rotter's claim that the external locus is not the negative one. Even with later revisions which differentiated those who react to a belief in external control with passivity from those who use the belief defensively, much of the use of Rotter's work has been implicitly to define individuals with a strong internal locus of control positively. Health education especially has used the notion of an internal locus of control to identify individuals who are most likely to react 'rationally" to health education. To risk overstating the case, the extensive use of Rotter's scale has meant that beliefs in fate and accident have only been studied as irrational lay beliefs; anachronistic and dysfunctional within the modern rational corpus of belief.

There has been little attempt to develop a social theory of the accidental which would occupy this space between broad sweep epidemiological explanations and the detailed individual background needed for a psychotherapeutic theory. Such a theory would examine how we construct the category and how it fits with our rational explanations of the world. The work in medical sociology has, naturally enough, focused on injuries. Figlio (1985), for example, has argued that the appearance of accidents is tied to the rise of capitalist relations of production. An accident, he claims, could not have happened before contractual working relationships developed. Intentional injuries, where compensation could be claimed if motivation could be proved, have a long history, but it was not until the Workers Compensation Act of 1897 that compensation could be claimed without proving malicious intent. A transitional stage was the fatality that could be compensated by a *deodand*,[2] which he describes as an ambiguous accident, unforeseen and not malicious, yet somehow implying intent' (Figlio 1985). Once the idea of negligence had entered master-servant law in the nineteenth century it became possible, Figlio argues, to conceive of the notion of an accidental injury.

There is a problem in this terminological slippage from 'accident' to 'industrial injury'. It becomes tautological to argue that industrial injuries could only happen once industrial modes of production were established, even if the processes and ideologies surrounding those connections may be fertile ground for research. Focusing exclusively on one kind of accident risks begging the question, 'why were such injuries categorized as accidents specifically?'. What Figlio has admirably charted is the inclusion of industrial injuries into the larger group of misfortunes that we call accidents and, in some senses, the inclusion of negligence as a culpable act. These new departures do not, though, appear to have significantly shifted the place accidents have had in our classificatory framework since the end of the seventeenth century. We still distinguish acts for which no blame can reasonably be apportioned from others, even if the former category has been shrunk by the removal of negligent acts. We are no nearer a social theory of how accidents interact with rationalist explanation.

A more useful approach for this purpose might be that adopted by Cornwell (1984) in her study of 'public' and 'private' lay accounts of illness. Cornwell noted that ideas of fate, destiny and luck were particularly salient to her respondents as explanatory variables in illnesses which were seen to be morally unambiguous, such as accidents (Cornwell 1984:214). The persistence of ideas about destiny are seen here to be evidence of the incomplete penetration of medical rationality as a dominant discourse of health and disease. Although health was seen as a moral category with 'good health being a morally worthy state and illness being discreditable' (Cornwell 1984:187), there were health problems which were externally caused and therefore incurred no moral scrutiny. Although Cornwell is not explicitly concerned with accidents as misfortunes, but as ambiguous categories of ill health, this is an important departure, as the features which characterize accidents are examined as ways of understanding the world, rather than misconceptions.

Within this account of lay beliefs though, accidents are still seen as marginal to ideas about health and illness, which are the central concerns of medical sociology. Perhaps the reason lies in the dependence of medical sociology on the structures of Western medicine: we have no 'sociology of misfortune' which might include an examination of the accidental, but a 'sociology of health and illness' within which the accident occupies a similar, peripheral place as it does in nosology. The dominance of medicine over the substantive concerns of sociology cannot, though, be held solely to blame for the relative neglect of the accident. The main traditions of social theory, like those of science, have provided little incentive for study of the events whose essential nature is not captured by rational explanation.[3]

The rational in the sociological tradition

The roots of rigorous study of the social lie in the establishment of rationalist thought and science. Sociology, as much as medicine, has been inevitably biased toward providing rational explanations for the behaviour of people and societies. Space does not permit more than a brief mention of two of the major traditions to illustrate this enduring incitement to rationalization, and to suggest the paradoxical place any adequate theory of accidents would have within these traditions.

Part of the Durkheimian project, for instance, was explicitly concerned with the uncovering of social laws to make sense of social behaviour. In *The Rules of Sociological Method* Durkheim claimed:

> our principal objective is to extend scientific rationalism to human behaviour. It can be shown that the behaviour of the past, when analysed, can be reduced to relationships of cause and effect (Durkheim 1950:xxxix).

Such an approach has no room for the capricious accident, which can furnish no useful data for such a project. Although *The Rules of Sociological Method* only represent a small, and perhaps atypical, part of Durkheim's work, the sentiments expressed here have been of enduring import to the positivist tradition. Durkheim's sociology here and in *Suicide* (Durkheim 1963) relied on the kind of statistics reviewed above that were well established by the end of the nineteenth century. Social facts, he claimed, are:

> represented with considerable exactness by the rates of births, marriages and suicides that is by the number obtained by dividing the average annual total of marriages, births and suicides by the number of persons whose ages lie in the range in which births, marriages and suicides occur. Since each of these figures contains all the individual cases indiscriminately, the individual circumstances which might have had a share in the production of the phenomena are neutralised and, consequently, do contribute to its determination. The average, then, expresses a certain state of the group mind (Durkheim 1950:8).

The accidental is to be 'neutralized'. Some apparently accidental events will only reveal their structured causes after appropriate investigation:

> facts most arbitrary in appearances will come to present, after more attentive observation, qualities of consistency and regularity that are symptomatic of their objectivity (Durkheim 1950:28).

Those that do not reveal such objectivity of observation are presumably not worthy of further investigation.

The Weberian tradition in sociology, which could perhaps more reasonably be expected to provide a theoretical starting point to the study of accidents, has also prioritized the rational. One of Weber's main concerns was the process of modernization, a process which is seen here to have both produced and denigrated the accidental as explanatory theory. However, in Weber's analysis, such modes of explanation are seen to be specifically a feature of pre-modern life. Giddens, for instance, quotes Weber on the comparison of modern capitalist society which is 'rationalized on the basis of rigorous calculation, directed with foresight and caution' compared with traditional peasant economy which is 'orientated to the exploitation of political opportunities and irrational speculation' (Giddens 1971:127). Belief in accidental cause appears as anachronistic within a modern rational cosmology, an example of ritual action that does not constitute proper social action. Action is only properly social, Weber claimed, 'in so far as, by virtue of the subjective meaning attached to it by the acting individual (or individuals), it takes account of the behaviour of others, and is thereby orientated in its course' (Weber 1947, quoted by Keat and Urry 1975:145). Only action so orientated is seen as a fruitful object of sociological inquiry.

In social anthropology there has been considerable debate about how to deal with apparently irrational explanations such as those reported by Evans-Pritchard (1976) among the Azande, who held witchcraft to be the cause of all misfortunes (Winch 1964, Hollis and Lukes 1982). Winch contested that Azande beliefs were not illogical but that definitions of rationality are social constructions that may have no relevance outside the Western cultural tradition: to evaluate Azande beliefs from the standpoint of Western science is a pointless ethnocentric exercise. Hollis and Lukes replied that rational scientific explanations were empirically testable, therefore superior, and that it was reasonable to judge other beliefs by them. The debate has been largely about other cultures: a rational ideological hegemony is assumed in the West.

The debate has also been largely marginal to mainstream sociology, which has focused on explanations of rational behaviour. We only observe the apparently irrational in order to discover the rational 'true' motivation, whether in the spirit of Durkheim, or that of structuralist theory, which attempts to uncover deep patternings underpinning superficial discontinuities. Behaviour or beliefs that remain unexplained are seen to be remnants of a pre-modern age. The traditional concerns of social theory have little space in which to explore the accident as constructed in everyday discourse and examine its place within a rational cosmology.

Rather than seeing the accident as epiphenomenal of a deeper rationality (as in the work of Freud) or as irrelevant in the grand sweep of social explanations, it might be helpful to focus on the accident itself as an ideal type of event which, by consensus, is morally neutral and which lies outside the bounds of rational explanation. To follow from Mary Douglas (1984) the boundary category may be the most informative about our classificatory systems, given the negotiation that surrounds candidate cases.

As an ideal type of event the accident is most usefully conceptualized as a boundary category in our classification of misfortune rather than as a category of illness or injury. Such an approach would have to explore three paradoxes surrounding accidents. First, they are seen to be universal ('accidents will happen'), but at the same time they are essentially unpredictable as unique events. Second they are morally neutral ('I didn't do it on purpose') but surrounded by moral inquiry. Third, and most enigmatically, they appear to be the product of a modern rational discourse, yet adherence to the accidental as explanation, marks the believer as anachronistic: whether superstitious, a peasant fatalist, or reprobate patient who refuses to believe they can control their own health and persists in belief in chance events.

The practical import of an analysis of the accidental in life lies perhaps in the implications for the victims of accidents. Freud is quite clear on the implications of his view for treatment of the sufferer:

> When a member of my family complains that he or she has bitten his tongue, bruised her finger, and so on, instead of the expected sympathy I put the question, 'why did you do that?' (Freud 1938:131).

It has been suggested that, as a society, we are becoming less capable of accepting uncertainty and risk (Fox 1980). How we construct accidental happenings is integral to our ideas of risk and uncertainty and coping with a world that we believe ought to behave in an ordered and regular fashion: accidents are disturbing in that they are a reminder of the limits to this rational cosmology.

As an explanation of misfortune, the accidental appears to have emerged at the end of the seventeenth century, with the advent of rationality. It is suggested that an adequate social theory of accidents would take this origin and the paradoxes surrounding accidental events as a starting point, rather than assuming that recourse to the accidental as explanation belongs to a pre-modern world, or that our common sense beliefs are simply mistaken. This paper has briefly raised some problems in the development of such a theory, given that accidents are precisely those events that many of the available theoretical approaches have dismissed as unworthy of attention or inexplicable.

Acknowledgements

I would like to thank David Armstrong for his helpful comments and advice and the Department of General Practice and Primary Care, King's College School of Medicine, for time and financial support for a larger study of which this is part.

Notes

1 For example: 'Accidents are by no means random occurrences' (Child Accident Prevention Trust 1989:16) or 'It is unfortunate that the word "accident" tends to imply an event which is unpredictable and therefore unpreventable, but of course accidents are as capable of analysis as any other event' (Jackson 1977:4).

2 In old English law, deodands were forfeits paid to the Crown as 'gifts to God' in cases of death by misadventure, to the value of the object (or sometimes animal) deemed to have caused the death. They were only formally abolished in 1846, and were collected as late as 1825. For examples of medieval fatalities which resulted in such forfeits see Sharpe (1913).

3 There have been attempts to theorize non-linear causal patterns (such as coincidences), most notably by Jung (1955) and by the Austrian biologist Paul Kammerer, who proposed a 'law of seriality' to complement scientific laws of cause and effect, earlier in the century (Koestler 1975). However these ideas never gained much credibility within the scientific community and have had little discernible impact. Koestler suggests that Kammerer's publication of his views may have cost him a professorship, despite his reputation at the time for his experimental work.

References

British Medical Association (1990), *The BMA Guide to Living with Risk*, Harmondsworth: Penguin.

Calnan, M. (1982), 'The hospital accident and emergency department: what is its role?', *Journal of Social Policy*, **11**:483–503.

Child Accident Prevention Trust (1989), *Basic Principles of Child Accident Prevention*, London: Child Accident Prevention Trust.

Cornwell, J. (1984), 'Health – a coincidence? Accounts of health and illness in a working class community: the case of Bethnal Green' (Ph.D. thesis), University of London.

Douglas, M. (1984), *Purity and Danger: an analysis of the concepts of pollution and taboo*, London: Ark.

Durkheim, E. (1950), *The Rules of Sociological Method* (trans. S.A. Solouay and J.M. Mueller), Glencoe: Free Press.

Durkheim, E. (1963), *Suicide: a study in sociology* (trans. J. Spaulding and G. Simpson), London: Routledge and Kegan Paul.

Evans-Pritchard, E.E. (1976), *Witchcraft, Oracles and Magic Among the Azande*, Oxford: Clarendon Press.

Figlio, K. (1985), 'What is an accident?', in Weindling, P. (ed.), *The Social History of Occupational Health*, London: Croom Helm.

Fo, D. (1980), *The Accidental Death of an Anarchist: A Farce* (adapted by Gavin Richards), London: Pluto.

Fox, R. (1980), 'The evolution of medical uncertainty', *Milbank Memorial Fund Quarterly/Health and Society*, **58**: 1–49.

Freud, S. (1938), *The Psychopathology of Everyday Life*, Harmondsworth: Penguin.
Giddens, A. (1971), *Capitalism and Modern Social Theory: An analysis of the writings of Marx, Durkheim and Weber*, Cambridge: Cambridge University Press.
Graunt, J. (1662), *Natural and Political Observations Mentioned in a Following Index, and made upon the Bills of Mortality*, London.
Green, J. (1992), 'The medico-legal production of fatal accidents', *Sociology of Health and Illness* (forthcoming).
Hacking, I. (1975), *The Emergence of Probability*, Cambridge: Cambridge University Press.
Hollis, M. and Lukes, S. (1982), *Rationality and Relativism*, Oxford: Basil Blackman.
Jackson, R. M. (1977), *Children, the Environment and Accidents*, London: Pitman Medical.
Jung, C. G. (1955), *Synchronicity: An acausal connecting principle* (trans. R.F.C. Hull), London: Routledge and Kegan Paul.
Keat, R. and Urry, J. (1975), *Social Theory as Science*, London: Routledge and Kegan Paul.
Koestler, A. (1975), *The Case of the Midwife Toad*, London: Pan Books Ltd.
Lodge, T. (1603), *A Treatise of the Plague, Containing the Nature, Signes and Accidents of the Same*, London: E. White and N. L.
Perrow, C. (1984), *Normal Accidents: Living with high risk technologies*, New York: Basic Books.
Petty, W. (1699), *Several Essays in Political Arithmetick*, London: R. Clavel and H. Mortlock.
Registrar-General (1956), *Annual Report of Births, Marriages and Deaths*, London: HMSO.
Rotter, J. (1966), 'Generalised expectancies for internal versus external control of reinforcements', *Psychological Monographs*, **80**: 1–28.
Rotter, J. (1982), *The Development and Application of Social Learning Theory*, New York: Praeger.
Sharpe, R. (1913), *Calender of Coroner's Rolls of the City of London*, London: Richard Clay and Sons.
Simpson, K. and Knight, B. (1985), *Forensic Medicine*, 9th edn, London: Edward Arnold.
Tombs, S. (1991), 'Injury and ill health in the chemical industry: decentering the accident prone victim', *Industrial Crisis Quarterly*, **5**: 59–75.
Ward, S. (1622), *Woe to Drunkards: A sermon*, London: John Marriott and John Grismand.
Winch, P. (1964), 'Understanding a primitive culture', *American Philosophical Quarterly*, **1**: 307–24.
World Health Organisation (1977), *Manual of the International Statistical Classification of Diseases, Injuries and Causes of Death*, Geneva: WHO.

4 The idea of prevention: A critical review

Richard Freeman

Introduction

Prevention is a widespread interest in social policy. It is of particular concern in attempts to confront crime and disease, and of particularly acute concern in some of the border regions between the two, such as drug use and child abuse. In broader terms, the idea of prevention may be thought of as being central to the welfare state.[1] In this light, the relative paucity of social scientific writing on the subject of prevention appears paradoxical. While there are some reviews of prevention in particular fields[2] and, beyond these, there is in each field a large and diverse literature which may be seen as implicitly related to prevention, nowhere is there an attempt to formulate a general understanding of prevention as a social policy goal. The existing literature is also very much of the applied, practice-oriented kind. It is concerned with models, strategies, frameworks and approaches; it lacks a critical reflexivity . No-one has done for prevention, for example, what Tawney attempted to do for equality (Tawney 1938), Rawls for justice (Rawls 1972) or Gough and Doyal (1991) for need.

It might be argued that prevention is not a concept of this order. Perhaps it is rather more of a secondary construct, a policy aim rather than a grand philosophical design. Yet there is literature on coordination, for example, variously described as 'collaboration', 'joint approaches' and 'joint planning' (Central Policy Review Staff 1975; Challis, Fuller, Henwood, Klein, Plowden, Webb, Whittingham and Wistow 1988; McGrath 1988; Nocon 1989 and Audit Commission 1986), and likewise on discretion (Adler and Asquith 1980), on participation (Richardson 1983), on administration (Dunsire 1973) and on accountability (Day and Klein 1987). The assumption in each case is that there is something essential to the theme, something generic, and that experience in one field may be illuminating in another.

It may be that part of the problem in talking about prevention in similar terms is that it belongs to neither of these two conceptual orders. It seems to occupy a middle ground: it is neither a grand philosophical design nor a simple requirement of policy, but it contains elements of both. At the same time, it may be the seeming incontestability of the proposition that 'prevention is better than cure' (Billis 1981) that obviates any more rigorous analysis. Nevertheless, as Billis points out in relation to prevention in social work:

> the attempt to attach meaning to and implement a sloppy idea is not of mere philosophical interest but is dysfunctional for clients, staff and the community (Billis 1981:379).

In respect of crime prevention, too, Steven Box writes of a lack of fit between policies and theories, and of a 'virtual wasteland of criminological theorizing' (Box 1987:195–6).

In this chapter, I do three things. First, I review some of the problems inherent in conceptualizing prevention, in associating it with ideas about causation and in relating these ideas to practice. This goes some of the way to explaining the frustration expressed by the writers I have just mentioned. Secondly, I try to expose what I see as some of the limitations of the conventional modelling of strategies of prevention into primary, secondary and tertiary categories. I discuss a further distinction between individual and social strategies and place them in the context of wider perspectives on social welfare. Thirdly, I go on to sketch certain dimensions of the politics of prevention. The purpose is to find ways of explaining what has been identified as a gap between rhetoric and reality (Holtzman 1979) in respect of prevention. My central concern is with the health field, though I draw argument and inspiration from discussions of prevention in other areas of social policy such as crime and child care.

I want to begin, however, by asking what is essential to prevention. I suggest, for the moment, that the concept of prevention is an amalgam of two others: prediction and intervention, foresight and action. This formula is intended, to serve only as a heuristic device; it is meant to do no more than introduce and order a set of ideas. The question of prediction draws attention to the status of epidemiological knowledge, while the question of intervention draws together a set of ideas about the ways in which strategies of prevention are conceived. Each element can be shown to have unresolved, and usually unacknowledged political aspects. These form the basis of discussion in the second half of this chapter.

Prediction: problems of aetiology

Prediction and intervention imply an understanding, or at least a hypothesis, of cause; in social scientific terms, the idea of cause tends to reflect positivist thinking. In a textbook view of preventive medicine, for example, the author asserts that:

> Facts are the tools of preventive medicine – facts about causation (Muir Gray 1979:24)

and that, in turn, epidemiologists are its toolmakers (Muir Gray 1979). Epidemiological work, however, typically ignores the ideological assumptions which inform the process of perception and description on which its data are based.[3] Its argument deals less with 'facts' than with ideological constructs: one enterprising and self-critical epidemiologist has set out 'to show causal concepts to be as ephemeral as history' (Susser 1973:viii). In respect of the nineteenth century, Tesh has pointed to the relationship between theories of disease causation and current political ideologies (Tesh 1982).

While the prevention of mental illness has been referred to as 'the possible science' (Bloom 1981), the key to the issue, perhaps, lies in David Billis's recognition of the fallacy of translating ideas of cause from the physical to the social world (Billis 1981). This involves not merely the application of an idea, but a wholesale transition between natural and social scientific epistemologies.[4] The social sciences, and the welfare professionals whose activity is based on them, have proved more successful at predicting the experience of populations than of individuals. This is in the nature of social statistics, which deals in groups and trends, in probable rather than actual experience. Social statistics lies at the heart of social epidemiology and some forms of criminology; in combination, it is this orientation to populations and probabilities which makes for controversy over the legitimacy of preventive interventions in respect of actual individuals.

The difficulties of the search for causes can be put in the following way. There are inherent problems in translating findings between different levels of social organization. As Susser explains, the circumstances of social class, for example, may be a direct cause of the variation in mortality rates between classes but are only an indirect cause of the mortality of the individual (Susser 1973).[5] In all epidemiological research individual deviations will be found from the normal experience of the group. There are, in consequence, serious ethical problems related to medical and social interventions based on the kind of information which epidemiology provides.[6] Decisions must be made, for instance, about the relative interests of the individual and the group. Decisions of this kind are ultimately political, ideological rather than 'scientific'. In this way, the statistical phenomenon opens an epistemological space in which the politics of social intervention (here 'prevention') are played out. In the case of child abuse, for example:

> research has failed to provide the results whereby abuse can be predicted and identified with any precision, so that health and welfare practitioners are being asked to do a job for which the basic tools do not exist ... failure to prevent child abuse has provided the opportunity for the New Right to restructure child care policy according to a more controlling, prescribed view of state welfare and a more pessimistic view of human nature (Parton 1985:149–50).

At the epistemological level prevention is, in short, an uncertain and contested field. It occupies an uncomfortable position at the intersection of social and medical science. Meanwhile, two further points put this critique into context. First, preventive intervention may be effective even in the absence of a proper understanding of cause. There are large areas of public health which are not based on a full knowledge of pathogenesis and aetiology (Tuomilehto and Puska 1987, Arnold and Banta 1977). To repeat two much-quoted examples, John Snow took the handle off the Broad Street pump more than thirty years before Pasteur's substantiation of germ theory, and sailors knew that lime juice prevented scurvy long before biochemists knew about vitamins. Secondly, the relationship between social science and public policy is a much less than straightforward one:

> ... policy makers choose selectively among available research to justify the decisions they make about the strategies they choose to label preventive. Conclusive data on the etiology of chronic disease would not dramatically change the character of the contemporary debate over prevention (Taylor 1982:40).

Intervention: problems of strategy

The policies and practices collectively labelled 'preventive' are normally divided into primary, secondary and tertiary forms. This categorization was first developed in the health field (Hardiker, Exton and Barker 1989) and has since been applied to others. Primary prevention refers to action designed to prevent the occurrence of disease: secondary prevention to that designed to prevent its worsening; tertiary prevention to that designed to prevent its debilitating effects. The introduction to the government's 1977 white paper *Prevention and Health* (DHSS 1977) sets out the order and range of preventive activity, and is worth quoting at length:

> Prevention in relation to health is either an attempt to prevent disease or disability before it occurs (primary prevention), the early detection and treatment of conditions with a view to returning the patient to normal health (secondary prevention), or the continuing treatment of disease or disability to avoid needless progression or complications (tertiary prevention).
>
> Prevention permeates virtually all aspects of the health services, not simply those which are normally regarded as mainly or wholly preventive. Moreover, a great deal of vital preventive activity takes place outside the National Health

> Service in such fields as education, housing, transport, employment, social services and environmental planning; and there are yet other fields at both national and local level which also offer important opportunities for prevention, for example, in the areas of taxation, prices and consumer protection, food hygiene and the provision of leisure and recreational facilities (DHSS 1977:82).

According to this general understanding, 'prevention' seems to embrace the entire realm of medical and health-related action. Perhaps inevitably, however, the inclusive approach masks significant conflicts and disagreements which take place behind this consensus. The conceptualization of successive orders or prevention points to the importance of boundaries between prevention, treatment and care, and between health services and other aspects of public policy.

The diversity of understanding which emerges is based on competing interests and perspectives. Williamson and Danaher, for example, interested in self care, develop a purist conception which promotes individual action by excluding medicine:

> Only primary prevention is truly preventive in nature; the other two are mainly concerned with treatment (early treatment and adequate treatment respectively) (Williamson and Danaher 1978:110).

Smith, by contrast, writes from a medical perspective, which tends to exclude the possibility of properly preventive, that is non-curative or non-medical action:

> For many diseases of slow, inevitable and insidious onset, the distinction between prevention and intervention is impossible to make. In fact, medicine has become interventive rather than specifically preventive or curative since it is concerned to intervene at the most effective point in the chain of causally related processes which comprise the development, onset and reduction of the disease (Smith 1968:209).

In context, this pair of quotations neatly expresses the unfocused nature of the debate. It repeats the pattern found in the general literature; the combined effect is that of a cleverly constructed *trompe l'oeil.* From one perspective, that of the conventional categorization, 'everything is prevention'; from another, that of Williamson and Danaher, 'only primary prevention is really preventive'; from another, Smith's 'there is not really any such thing as prevention'.

Primary, secondary and tertiary interventions reflect what is taken to be the natural course of pathology from onset through to death. What might be described as the linear conception of prevention is derived from the medical or disease model of health and illness.[7] Conventionally, the weight of attention is to primary and secondary strategies, while tertiary forms of prevention may be victim to a consensus, perhaps both lay and professional, that they are 'not really preventive'. In consequence, much of the conflict over prevention is located on the boundary between these primary and secondary forms: it sets arguments for prevention through public health and health promotion against those for prevention by medicine (screening) and it includes competition for resources between the two.

The literature on prevention also reveals a different axis of definition. Strategies of prevention can be remodelled according to the focus of their intent, which is either 'individual' or 'social'. This axis crosses the primary-secondary-tertiary categorization. It is not, however, a neat bisection; primary prevention tends to require institutional or social input, while secondary prevention attends more usually to the individual (Caplan 1966; Morris 1975). The result is best expressed in the following reformulation: primary prevention aims at the adaptation of circumstance to individual need; secondary prevention at the adaptation of the individual to circumstance (Vobruba 1983).[8] Strategies which seek to prevent while upholding prevailing economic and social structures may be distinguished from those which seek to prevent by altering those structures themselves.

This sort of definition implies a simplistic conception of the relationship between individual and social spheres as essentially dichotomous: they are unable to show the interactions between the two. In practice, too, the distinction between individual and social is difficult to sustain. Some of the more widely-used preventive programmes, for example, such as screening and health education, are targeted at the mass. It is arguable that these should properly be thought of as individual forms, from which they can be distinguished only by the quantity of their focus and not by any qualitative conception of the significance of the 'social' (Karmaus 1981). It is rarely clear what is meant by the 'social'; it might equally apply to culture (social behaviour) and to the economy (modes of production and employment) as to the urban environment (housing, transport). In the minds of many advocates of prevention, it refers loosely to all of these.[9]

The important of this construction of individual and social is that it indicates a link between prevention and political values. These are mediated, in turn, by ideas about welfare. It has recently been argued that:

> it is consideration of the value base underlying the intervention which helps towards greater conceptual clarity about preventive work (Hardiker, Exton and Barker 1991:355).

One way of doing this is to relate primary, secondary and tertiary categories of prevention to institutional, residual and developmental models of welfare (see Hardiker, Exton and Barker 1991). In the health field, the institutional model of welfare tends to represent universalist concern for whole populations and the belief that planned intervention by the state will be effective. Public health and primary prevention, then, fit best with this model. In a residual system, prevention in health is left to medicine to manage as it sees fit and as demands are made upon it. Most attention is paid to the secondary level of prevention, to screening and some health education. Distinctions between primary, secondary and tertiary prevention are least clear in the developmental model; indeed, it may be that prevention is virtually excluded from this perspective.[10] It appears, therefore, that:

> the term *prevention* has little heuristic value unless it is interrogated in relation to the assumptions underlying the specific model of health care adopted (Hardiker, Exton and Barker 1989:23).[11]

The idea that strategies of prevention can be mapped onto models of welfare is valuable in that it adds weight to the argument that the formulation of preventive social policy may not be dependent, in the end, on technical understandings of the genesis of social problems. The argument is that prevention is more closely related to social values and political and institutional organization. The production of policy for prevention, both whether prevention becomes a policy interest in social welfare and in what form, is an effect of the interplay of ideologies and interests in any given field. It is from these that models of welfare are derived and which, in turn, reflect the distribution of power in society. They suggest further ways of exploring the dynamics of prevention in social policy.

Towards a political understanding of prevention

In the English tradition of political theory, prevention might best be thought of as representing a Fabian goal, reflecting a vision of social engineering and a faith in the capacity of the state to plan and to act. As such, of course, it is subject to liberal critiques (Anderson 1980). These address the dangers of the prospect of prevention being allowed to legitimate increasing intervention by the state, and they are usually combined with attacks on the welfare state per se. In short, prevention arouses the classic political problem of how to square individual freedom with the public good.[12] The introduction of the compulsory wearing of seat belts in the UK was entangled for a long time in the 1970s in just this kind of debate.

This conflict of individual freedoms with the public good might be expected to emerge as a clash between liberal and social democratic thought (Schuller 1982). In practice, however, what is interesting is the extent to which the rhetorical power of prevention is felt across the political spectrum. It seems to offer opportunities both for social engineering and for increased personal responsibility:

> To the left it is part of the progress towards a better society where the social ills of capitalism will be diminished, to the right prevention holds out the hope of reduced government intervention of concrete and effective policies (Allsop 1984:173).

The attraction may also be due in part to prevention's common sense appeal: being for prevention is the health political equivalent to being against sin. Beyond that, prevention is a political promise with a strong utopian aspect (Klages 1982): successful prevention might mean a world without disease, crime, unemployment or poverty. At the same time, many issues to do with prevention do not fall consistently into the ideological divide between left and right. They include those which, like sexual health, are ascribed to the sphere of the personal rather than the public and others, such as those having to do with gender and race, which established political thought has largely failed to take into account.

Meanwhile, prevention may be driven as much by economic as by political concerns. The point may be less that 'prevention is better than cure' than that 'a stitch in time saves nine'. The dream of cost containment in health care through investment in medical screening, for example, was held up in the United States in the 1960s and 70s by what has been characterized as 'a coalition of fiscal conservatives and social services liberals' (Stone 1986). More recent work on cost-benefit analysis, however, has indicated that there seem to be no financial gains to be derived from diagnostic screening (Russell 1986; Bryan 1990).[13] Interestingly, it may be the marginality of preventive work which makes it vulnerable to attempts at evaluation:[14]

> Almost alone within the police service, crime prevention officers are working in an environment where profits, costs, advertizing, public resources ... and political interests shape just about everything they do. Economics has a powerful influence on crime prevention (Love 1987, cit Grimshaw, Harvey and Pease 1988).

What are held up as economic considerations may well run against rather than in favour of prevention.

From the broader perspective of political economy, policy options for prevention may be foreclosed by the threat they present to economic and industrial interests. In its traditional form, public health may be compatible with classical economics, which provides for the regulation of externalities or 'public goods'. The state has a role in supporting the economy but, importantly, stops short of interfering in it. More critically, Marxists have argued that there is a class-based interest in policing over prevention in health (Breilh 1979; Labisch 1982) and that medicine under capitalism is necessarily oriented to the treatment of individuals rather than to the protection and maintenance of the public health (Renaud 1975; Navarro 1976;

Doyal 1979; Waitzkin 1989).[15] This provides some explanation for a pattern of selection of secondary and individual over primary and social strategies of prevention. Secondary forms accord well with the individualized consumption of health care, which is based on the doctor–patient relationship and is supported by commercial interest in technological and pharmacological innovation. The screening of populations, for example, can be constructed as a set of individual transactions. Similarly, in parallel with formal medical activity, health issues play a prominent role in the contemporary marketing of lifestyle. Prevention at the primary level of public health, however, offers no basis for cost accounting and cannot be marketed in a comparable way (Labisch 1982).

The relationship between the welfare professions and the idea of prevention is an uneasy one. Prevention represents only a marginal aspect of the dominant thinking and work practice of the core institutions in health and welfare. For the police, for example, 'enforcing the law' has always been accorded a higher priority than 'keeping the peace' (Heal 1983). For the doctor, the locus of both scientific and ethical interest is the sick patient and not the seemingly well. There are traditions both in medicine and in law which protect the individual from anything but *post factum* attention; perhaps more fundamentally, Labisch suggests that technocracies are organized to respond to symptoms of problems rather than to their causes (Labisch 1982). That said, there is reason to suggest that the professions take an instrumental view of prevention. Prevention may be used, for example, to capture and control demand for professional services.[16] Claims for 'prevention' may serve in fact to extend the realms of 'cure' and 'care'; preventive programmes increase the number of points of access to the individual. The Cumberlege Report of 1986 argued for an extended role for nurses in primary and community care which was partly based on a potential contribution to prevention (DHSS 1986). It is, perhaps, the lack of specificity inherent in the idea of prevention which makes it a vehicle of professional opportunism:

> The very broad scope of primary prevention raises questions about the boundaries of social work's permitted and required endeavours (Hardiker, Exton and Barker 1991:348).

Ordinarily, user groups in health care tend, almost by definition, to be concerned with issues of access and service quality. More widely, the coalescing of public pressure in response to a given issue is precisely that: it is usually reactive rather than proactive. Where public concern is expressed, it tends to be about hazards that kill numbers of people in a catastrophic fashion rather than in a chronic, one-at-a-time way (Fischoff, Hohenemser, Kasperson and Kates 1980). Nevertheless, concern for health is a significant platform in environmental lobbying, for example, levels of lead pollution in urban play areas and in the water supply. Black community groups are addressing among others the problems of sickle-cell anaemia and the mental health of young people (Knowles 1991).

Examples of resistance to preventive programmes, meanwhile, tend to support interpretations of preventive policy-making which refer to professional and organizational self-interest. Feminists have exposed and contested many of the ideological assumptions and material interests which underpin, for example, both health education (Rodmell and Watt 1986) and new reproductive technologies (Stanworth 1987). Health education in relation to AIDS has provoked strong reactions, from gay men and drug users against the separation of 'innocent' from 'guilty' individuals with HIV; from women against representations of them being responsible for the prevention of heterosexual transmission of HIV, and against the pilot testing of screening programmes on pregnant women (see Chapter 11); and from groups of people with HIV against fatalistic representations of their experience and against the roles of government and industry in the production, licensing and marketing of AIDS-related pharmaceuticals (Carter and Watney 1989; Patton 1990). Developments like these may represent 'new politics' of prevention:

> The new political line-up often has labor, women's groups, parents' groups, advocates for women and children, and a new breed of 'ethics watchers' pitted against medical researchers, public health advocates and industry. The new rhetoric emphasizes the paternalism of prevention measures, the dangers of coercion in the name of better health, and the protection of individual rights. The resistance uses the strategies of civil rights politics: identifying groups as minorities or victims of discrimination and oppression, mobilizing around these identities, and pursuing political objectives through the courts (Stone 1986:672).

Prevention as a problem of governance

Briefly summarizing what has been said so far, the development of strategies of prevention (whether primary, secondary or tertiary, and whether individual or social) will be structured by scientific epistemologies. They will be shaped, in turn, by political values and professional ideologies and interests. They will be judged by their immediate resource implications, and by the extent to which they will impinge on existing material interests. They may be proposed and implemented both in response to and in spite of political pressure of varying kinds. Taken together, these factors suggest a possible map of the 'politics of prevention'.[17]

Taking further the two sets of ideas I have begun to outline here; those which have to do with social prediction and those which have to do with strategies of intervention, discussions of prevention might be informed by critiques of the relationship between knowledge and power (Foucault 1980). These are inextricable, shaping and being shaped by the activities of the state and the professions respectively; they are bound together in the concept of surveillance. This kind of theoretical approach is evident in critiques of containing institutions, such as the prison and the asylum (Foucault 1979), of the relationship between social welfare and social control (Sachße and Tennstedt 1986), and it might be said to culminate in warnings of the need to 'prevent prevention' (de Leonardis and Mauri 1983). The perspective has been enhanced by the development of a sociology of the body, which conceives of changing relationships between state and society and between individuals being mediated through ideas about health, disease and the body (Armstrong 1983; Turner 1982).

In a complementary fashion, prevention might also be construed as a problem of governance. Preventive policy-making will be structured by the kinds of pressure which particular issues exert on the political system (Vobruba 1983). Preventive policy, I want to suggest, is neither solely nor directly expressive of ideology and interest; it has a particular role or function in the policy-making process. In so far as issues constitute challenges to the social and political order, preventive policy-making represents not so much the means by which challenges themselves are confronted, but the means by which, in spite of such challenges, policy-making authority is maintained and reproduced. In this sense, prevention is an aspect of the management of social problems, not of their elimination.

This idea is strengthened by exploring the disjuncture between the ideology and practice of prevention. Some social work projects, for example, may act at more than one level of prevention: they may understand the issue with which they deal at one level – they may hold a 'primary' understanding of crime as being related to social circumstance, say – but may then order their intervention at another, say the secondary level of the individual 'at risk' (Hardiker, Exton and Barker 1989). David Stone notes, more generally, that preventive activities in health receive official support in inverse relation to their probable effectiveness, a pattern which he describes as 'upside down prevention' (Stone 1989).[18] This pattern is explicable as the product of political calculation: socio-environmental change is costly, radical and unpredictable, whereas health education is cheap, uncontroversial and safe. This strategy ensures that policy makers and politicians can claim credit for any eventual success, while target populations can be blamed for any failure to respond (Stone 1989). An appreciation of the context of policy-making serves to strengthen this argument. Preventive policy tends to formulate and implement at a point of crisis in the evolution of a social problem.[19] Active policy for prevention may be a response to immediate pressures, while primary strategies can expect, at best, only long-term results. Equally problematically for the politician, results will be difficult to attribute to any specific policy intervention. Immediate pressure on the state is best relieved, therefore, by displacing responsibility for prevention in the direction of the individual, that is into the secondary realm (Vobruba 1983).

This hypothesis can be refined. Primary interventions are likely to disturb interests which have functional value to the state itself. By embarking on primary strategies, then, the state merely exacerbates its burden of problems. Primary strategy is not calculated to relieve a political crisis. The dysfunctional or counterproductive effects of secondary strategies, on the other hand, emerge only in the long term. Their success is dependent on the compliance of individuals, and this can be expected to wane (Vobruba 1983). Secondary prevention, then, tends to reproduce the crisis which it was designed to resolve. This, in turn, is met either by further forms of secondary prevention or by curative and compensatory measures. The effect is that responsibility for the preventive maintenance and reproduction of the political, economic and social system is displaced onto the individual. As Vobruba puts it:

> From their position counter to the system, individuals are asked to behave in such a way as to keep it intact (Vobruba 1983:34).[20]

One of the key functions of a policy of prevention, then, is to reproduce the willingness of individuals, firstly to collude in the ideological reconstruction of social problems as individual problems and then, secondly, to cooperate, as individuals, in prescribed solutions to those problems.[21]

In a similar way, though painting a much broader canvas, Schülein sees prevention as the conduit of social development (Schülein 1983). The more complicated a society, he argues, the more complicated is its future. Social development becomes more contingent on internal economic, social, political and technological processes and less so on external factors such as the weather, changing seasons and natural disaster. At the same time, it becomes less possible to predict the future on the basis of tradition and experience.[22] Consequently demands for, and attempts at, prevention become increasingly complicated. The goal of prevention shifts from the consensual protection of the social system from external threat to maintenance of that system in the face of its own internal, dysfunctional effects.[23] Different ideologies and interests within a social system will compete to construct different futures for that system. Essentially, then, prevention represents the means of production of particular futures (Schülein 1983).

This underscores the extent to which prevention is a normative process (Schülein 1983), which can be taken to operate at a number of levels. Definitions of crime, for example, as of disease and child abuse, are value-laden. Social intervention, then, when it is justified as prevention, almost inevitably tends to the normative: we prevent what is 'bad' in order to promote what is 'good'.[24] MacIntyre describes the promotion and prevention of childbirth as two potentially contradictory themes in gynaecological work, and the ways in which they are managed according to clinicians' understandings of 'normal family life' (MacIntyre 1976).[25] Similarly, in respect of crime prevention, Cohen suggests that the state uses 'assimilative' control, in the attempt to normalize or absorb deviants (Cohen 1985).

At the same time, prevention questions the legitimacy of routine social processes. These may be pathogenic, but they are also constitutive of the prevailing economic and social order and, by implication, of its forms of government. There is therefore an inherent tendency for prevention to question the bases of the authority through which it is promulgated (Schülein 1983). In consequence of this, prevention brushes against what we might call the self-interest of the state (Vobruba 1983).

> (Prevention) now aims no longer at the individual potential bearer of social policy problems, but at the preservation of the functioning of the social state. Prevention now no longer refers to the treatment of the causes (whether subjective or objective) of social policy problems, but to the shielding of social policy from problems; not the prevention of the emergence of problems, but the prevention of their being registered as such (Vobruba 1983:39).[26]

Discussion earlier in this chapter dealt broadly with the potential for instrumental uses of prevention. These refer to the possible extensions of power by means of policies described and advocated as preventive. To them must now be added that of the capacity of prevention for social and political destabilization, that is for the undermining of power.

Prevention as rhetoric and symbol

For government, then, prevention represents something of a double-bind. Government may gain much by constructing social phenomena as social problems, indeed its legitimacy may be derived in part from its advocacy of the prevention of certain social problems, but it also risks much by then intervening against those problems with any consequence. Preventive policy-making must seek to do two things: it must acknowledge the pathogenic as a problem and must avoid making a commitment to action. Prevention walks a tightrope between action and inaction, between promising and pretending. It is this tension which determines the symbolic nature of preventive policy-making.

Edelman's work offers further insight into this aspect of the problem (Edelman 1964, 1971, 1977). Though primarily interested in the extent to which participation in government is symbolic (Edelman 1964), Edelman is close to a consideration of policy when he writes of symbolic acts, suggesting that:

political acts are both instrumental and expressive (Edelman 1964:12).[27]

This is to say that all policy-making is symbolic in some way; it has meaning as well as function and purpose. Symbolic policy is that of which the prime purpose is to be expressive, is to confer and to communicate meaning rather than to achieve specified goals. It is instrumental only in so far as it is expressive. The goal of symbolic policy is achieved when ideas and understandings are represented to and accepted by those at whom it is addressed. Crudely, its aim is not to change the world but to change the ways in which the world is understood. The essential aspect of the symbol, furthermore, is that it is a representation of reality, not a reflection of reality. Symbols are derived from mediated perceptions of the world, not from the world as it is. This suggests, in turn, that a certain remoteness will be characteristic of symbolic policy-making. As Edelman puts it,

> What matters is remoteness, not content. Thus an actual enemy attack or an earthquake has always given men greater confidence and made them act more effectively, for this is something close and concrete with which to cope. It is reports of feints and omens that bring expressive acts with little instrumental value (Edelman 1964:13–14).

Engagement in symbolic politics, then, affords societies and polities a way of seeking collective reassurance against continuing threats, the causes or even precise nature of which remain unclear or unknown.

The importance of Edelman's work to this discussion is that it offers a means of transition between the theoretical realm of the sociologists (Schülein 1983; Vobruba 1983) and the realm of practical politics. The remoteness which is integral to the symbol is translated into an absence of detail in policy formulation brought about both by the remoteness of political actors from the reality with which they intend (or pretend) to engage, and by a concern to avoid binding and specific commitments to action. This implies that one of the more important criteria by which to judge preventive policy will be the extent of its rhetorical quality.

This chapter began by alluding to a gap, as regards prevention, between rhetoric and reality. The rhetoric of prevention contains an identifiable paradox: nowhere is there agreement about quite what prevention is, while everywhere there is agreement that it is a good thing. Prevention is a good thing, we may assume, because social problems are bad things. Social problems, moreover, arouse conflicts of ideology, interest and so of conceptual understanding. The idea of prevention, on the other hand, is very soothing. That means, in practice, that it is essential to preventive policy-making that it be vague. Policies and positional statements invariably promise 'more prevention' without declaring either what it is or, to coin a phrase, whose business it should be.[28] This is, in part, because technical knowledge is insufficient to allow such statements to be any more precise; as the first part of this chapter set out to show, this is almost inevitably so. In large part too, however, the vagueness of preventive policy-making is deliberate. Prevention must straddle ambiguities both of understanding and of political intent:

> Prevention has always had an ambiguous status in health care politics. It is at once the darling of health care reformers and the stepchild of health care professionals and politicians. Everyone is in favor of the idea of prevention – stopping disease and injury before they happen – but few want to stake a career on such an uncertain business or invest public funds in preventive measures (Stone 1986:671).

The rhetoric in which it is conceived is essential both to the politicization of social problems as matters for public policy, and to their eventual depoliticization.

This chapter has been an attempt to gain purchase on the idea of prevention in a number of different ways. It began by explaining that positivistic applied social science is unlikely to resolve conflicts over the appropriateness of preventive policies. Indeed, epidemiological and criminological research will often serve to highlight and even exacerbate those differences. At the same time, the conflation of the various functions of prevention, cure and care into one all-embracing model of health-related activity reveals a potential for the displacement of prevention into areas more usually thought of as proper to medicine or nursing. In effect, uncertainty about distinctions between primary, secondary and tertiary prevention is not merely *diffused into* but *distracts from* uncertainty about boundaries between prevention, treatment and care. Within the health sector, these are organizational, professional and status boundaries between public health, medicine and nursing, between professional and lay care, and between the health sector and the realm of government. What is important about understandings and definitions of prevention is not just that they are unclear, but that they are malleable. The language of prevention offers a means of eroding, crossing and redefining boundaries both between health-related activities and between those activities and others.

This analysis suggests that prevention may be used in struggles for authority within and between different parts of the health sector and between the health sector and its social and political environment. Put simply, prevention is a political construct. A possible relationship between models of prevention and models of welfare confirms that impression, begging in turn questions about the material and ideological forces which lie behind the making of social policy. Some of those forces can be quickly outlined: they include aspects of prevailing political thought, of the economics of prevention, of the interests in prevention of the welfare professions, and of the ways in which preventive programmes are demanded, experienced and resisted by individuals and groups.

One further train of thought, however, deserves particular emphasis. It engenders a set of related ideas about prevention and government, and is derived from the perception that preventive policy, while it may be a response to a variety of interests in pluralist society, is also an act of government. Preventive policy reflects two dominant concerns on the part of the state: these relate, perhaps self-evidently, to the security of its subjects, but they also include the security of security (Vobruba 1983), that is of the prevailing social and political order and of the place of government and the state within it. This implies that the state will seek to relieve the pressure put upon it by displacing responsibility for prevention from the primary to the secondary, that is from the institutional to the individual realm. It suggests, also, that much policy-making for prevention will be symbolic, characterized by a high level of rhetoric and by a low level of specific commitment to action.

Notes

1 In practice, however, the provision of welfare may be held to serve mitigating or compensatory rather than strictly preventive functions in relation to need.

2 For examples, see Rapoport (1961) and Billis (1981) on social work in the US and the UK respectively, Herriger (1986) on juvenile delinquency (in German), Dorn and South (1987) on drug use, Abbott (1971) on racial discrimination, Bloom (1981) and Newton (1988) on mental illness, again in the US and the UK respectively, and Holman (1988) and Hardiker, Exton and Barker (1989 and 1991) on prevention in child care. Hardiker, Exton and Barker (1989) also provide useful literature reviews of prevention in crime and health. In criminology, the literature on prevention is more extensive, though it seems to be without a general introduction; Clarke (1980), Heal (1983) and Hope and Murphy (1983) are useful in this respect.

3 For a critique of epidemiology on the grounds of its methodological positivism, see Chalmers (1982). Meanwhile, the field has serious shortcomings even on its own terms. It has been pointed out that most of what we know as the 'causes' of psychiatric disorder, for instance, are neither necessary, nor sufficient, nor even very specific (Cassel 1976; Newton 1988:45).

4 For practical purposes, the distinction between natural and social scientific conceptions of cause parallels a different distinction between professional (or medical) and lay understandings. For studies of the lay perspective, see Pill and Stott (1982), Blaxter (1983) and Calnan (1987).

5 This is why epidemiological information about the origin of disease is expressed in terms of indirect causality, multiple causality, semicausality and risk factors. Moroney has an appropriate metaphor for the basic dilemma which this kind of imprecision reflects: 'A statistical analysis, properly conducted, is a delicate dissection of uncertainties, a surgery of suppositions. The

surgeon must guard carefully against false incisions with his scalpel. Very often he has to sew up the patient as inoperable' (Moroney 1951:3).

6 Working as a criminologist, Leslie Wilkins is quite clear about the ethical responsibility of the researcher: 'I have conducted research into the prediction of recidivism and see no objection to this, but I have not, and would not, carry out research aimed at predicting probable delinquency. No individual citizen (and this includes juveniles) while that individual retains his full quota of individual rights, should be placed in a position of risk of becoming a "false positive"' (Wilkins 1985:35).

7 For a useful discussion of the disease model in relation to the prevention of child abuse, see Parton (1985).

8 '... primäre (institutionelle) Prävention (läuft) auf Entlastungen der Subjekte von Systemzwängen hinaus ... Sekundäre Prävention legt den Subjekten Arrangements mit institutionellen Zwängen nahe ... (und) läuft somit auf flexible, problemabsorbierende Anpassung hinaus' (Vobruba 1983:29–30). ('Primary (institutional) prevention amounts to relieving the individual of systemic pressures to conform ... secondary prevention suggests adaptation to institutional pressures ... (and) so amounts to a pliable problem-absorbing conformism.')

9 For a useful discussion of the idea of the social in social policy, see Squires (1990).

10 The developmental approach is based on recognition of and response to the needs and demands of citizens, clients and service users. These groups, however, rarely express a demand for prevention. The self-help group, for example, typically holds rehabilitative aims (Trojan, Soden and Wallenczus 1990).

11 In fact, these authors intend their work to be applied in respect of child care, in which the fit between models of prevention and models of welfare would be the following: 'In a society where a developmental model holds sway, the ideal level of prevention is primary: improve social conditions so individuals do not need to become clients. Where an institutional view of welfare is predominant, preventive efforts are pitched at the secondary level: early intervention to prevent problems worsening. Under a residual value system, prevention typically takes place at the tertiary level: work with children and families in imminent danger of separation, often through court proceedings' (Hardiker, Exton and Barker 1991).

12 The phrase is that used in an historical account of the prevention of socially and sexually transmitted disease (Porter and Porter 1988:97); for broader discussions, see in particular Yarrow's discussion of Mill in relation to preventive health politics in Britain (Yarrow 1986), and Wikler (1978) on the US.

13 In light of this, continued commitment to screening may reveal a symbolic function of prevention in health policy, by which increased provision of certain forms of prevention serves to legitimate parallel attempts at cost containment through service reduction in other areas: see Bauch (1989) on the reform of German health care.

14 There is an engaging set of problems which have to do with the evaluation of preventive activity. There are obvious, though unusual, methodological diffi-

culties involved in measuring and explaining the non-occurrence of a specified event (Grimshaw, Harvey and Pease 1988).

15 See also Doyal and Epstein (1983) for a specific discussion of the politics of cancer prevention.

16 Stanley Reiser has described the way in which, on the replacement of fee-for-service payment by health insurance, multiphasic screening was used in the US to regulate the flow of patients into the medical delivery system (Reiser 1980).

17 The phrase is Rosemary Taylor's (Taylor 1982).

18 Compare Julian Tudor Hart's 'inverse care law' (Tudor Hart 1971).

19 For discussion of the AIDS example, see papers by Weeks (1989) and Berridge and Strong (1990). For a valuable introductory account of the theory of social problems, see Manning (1987).

20 'Aus ihrer Gegenlage zum System sollen die Subjekte so handeln, daß es intakt bleibt' (Vobruba 1983:34).

21 Crawford speaks bluntly of the ideology and politics of victim-blaming in health policy (Crawford 1977).

22 Aubert refers to what might be described as the anthropological perspective on prevention: 'On the whole, it seems to be the consensus of anthropological opinion that misfortunes and other events, are never explained by reference to a chance concept. On the contrary, primitive man's systems of belief usually give evidence of a very vigorous denial of chance and uncertainty, and a similarly desperate affirmation of their capacity to master the world by secular skills or by magic' (Aubert 1980:75); see also Klages (1982) and Ames and Janes (1987).

23 Abram de Swaan shows how social contradictions are resolved by being deferred to medical authority (Swaan 1990); most of the processes and activities to which he refers – the medicalization of daily life, early warnings of health risks and medical advice – belong to the realm of prevention.

24 In the words of the Canterbury Report on coronary heart disease, 'Prevention is essential. Prevention implies the *correction* of causes' (HEC 1984:1, emphasis added).

25 See also Graham (1979) on the normative pressure on mothers to take responsibility for child health.

26 'Sie zielt nun nicht mehr auf den einzelnen, den potentiellen Träger sozialpolitischer Probleme, sondern auf die Erhaltung der sozialstaatlichen Handlungsfähigkeit. Prävention bedeutet nun nicht mehr die Bearbeitung der (subjektiven oder objektiven) Ursachen sozialpolitischer Probleme, sondern die Abschirmung der Sozialpolitik von Problemen; nicht die Verhinderung von Problementstehung, sondern die Verhinderung von Problemanmeldung' (Vobruba 1983:39).

27 Edelman's argument develops earlier work by Ulf Himmelstrand (Himmelstrand 1960).

28 The title of the UK government's 1976 discussion document suggested that prevention in health was 'everybody's business' (DHSS 1976).

References

Abbott, S. (ed.) (1971), *The Prevention of Racial Discrimination in Britain*, Oxford: Oxford UP.

Abholz, H.-H. (1980), 'Möglichkeiten und Vernachlässigung von Prävention', in Deppe, H.-U. (ed.), *Vernachlässigte Gesundheit. Zum Verhältnis von Gesundheit, Staat, Gesellschaft in der Bundesrepublik Deutschland. Ein kritischer Überblick*, Köln: Kiepenhauer und Witsch.

Adler, M. and Asquith, S. (eds) (1980), *Discretion and Welfare*, London: Heinemann.

Allsop, J. (1984), *Health Policy and the National Health Service*, Harlow: Longman.

Ames, G.M. and Janes, C.R. (1987), 'Anthropology and prevention in North America', *Social Science and Medicine*, **25** (8): 921–2.

Anderson, D. (ed.) (1980), *The Ignorance of Social Intervention*, London: Croom Helm.

Armstrong, D. (1983), *The Political Anatomy of the Body*, Cambridge: Cambridge UP.

Arnold, C.B. and Banta, D.H. (1977), 'Introduction: lessons from the past, deliberations about the future', *Preventive Medicine*, **6**: 191–7.

Aubert, V. (1980), 'Chance in social affairs', in Dowie, J. and Lefrere, P. (eds), *Risk and Chance, Selected Readings*, Milton Keynes: Open University Press.

Audit Commission (1986), *Making a Reality of Community Care*, London: HMSO.

Bauch, J. (1989), 'Das Gesundheitsreformgesetz – ein Paradigma symbolischer Politik', *Medizin, Mensch, Gesellschaft*, **14**: 59–68.

Berridge, V. and Strong, P. (1990), 'AIDS policies in the UK: a preliminary analysis', in Fee, E. and Fox, D. (eds), *AIDS: Contemporary History*, Princeton: Princeton UP.

Billis, D. (1981), 'At risk of prevention', *Journal of Social Policy*, **10** (3): 367–79.

Blaxter, M. (1983), 'The causes of disease. Women talking', *Social Science and Medicine*, **17** (2): 59–69.

Bloom, M. (1981), *Primary Prevention: the Possible Science*, London: Prentice-Hall.

Box, S. (1987), *Recession, Crime and Punishment*, Basingstoke: Macmillan.

Breilh, J. (1979), 'Community medicine under imperialism: a new medical police?', *International Journal of Health Services*, **9** (1): 5–24.

Bryan, J. (1990), 'Prevention's high price', *The Guardian*, 26 January, 29.

Calnan, M. (1987), *Health and Illness – The lay perspective*, London: Tavistock.

Caplan, G. (1966), *Principles of Preventive Psychiatry*, London: Tavistock.

Carter, E. and Watney, S. (1989), *Taking Liberties. AIDS and cultural politics*, London: Serpent's Tail.

Cassel, J. (1976), 'The contribution of the social environment to host resistance', Fourth Wade Hampton Frost lecture, *American Journal of Epidemiology*, **104** (2): 107–23.

Central Policy Review Staff (1975), *A Joint Framework for Social Policies*, London: HMSO.

Challis, L., Fuller, S., Hemwood, M., Klein, R., Plowden, W., Webb, A., Whittingham, P. and Wistow, G. (1988), *Joint Approaches to Social Policy*, Cambridge: Cambridge UP.

Chalmers, A. (1982), 'Epidemiology and the scientific method', *International Journal of Health Services*, **12** (4): 659–66.
Clarke, R.V.G. (1980), '"Situational" crime prevention: theory and practice', *British Journal of Criminology*, **20** (2): 136–47.
Cohen, S. (1985), *Visions of Social Control. Crime, punishment and classification*, Cambridge: Polity.
Crawford, R. (1977), 'You are dangerous to your health: the ideology and politics of victim-blaming', *International Journal of Health Services*, **7** (4): 663–80.
Day, P. and Klein, R. (1987), *Accountabilities: Five public services*, London: Tavistock.
Department of Health and Social Security (DHSS) (1976), *Prevention and Health: Everybody's business. A reassessment of public and personal health*, London: HMSO.
DHSS (1977), *Prevention and Health*, EM 7047, London: HMSO.
DHSS (1986), *Neighbourhood Nursing – A Focus for Care. Report of the Community Nursing Review, chaired by Juliet Cumberlege*, London: HMSO.
Dorn, N. and South, N. (eds) (1987), *A Land Fit for Heroin? Drug policies, prevention and practice*, London: Macmillan.
Doyal, L. with Pennell, I. (1979), *The Political Economy of Health*, London: Pluto Press.
Doyal, L. and Epstein, S.S. (1983), *Cancer in Britain. The politics of prevention*, London: Pluto.
Dunsire, A. (1973), *Administration. The word and the science*, London: Martin Robertson.
Edelman, M. (1964), *The Symbolic Uses of Politics*, Urbana: University of Illinois Press.
Edelman, M. (1971), *Politics as Symbolic Action*, Chicago: Markham.
Edelman, M. (1977), *Political Language. Words that succeed and policies that fail*, New York: Academic Press.
Fischhoff, B. Hohenemeer, C., Kasperson, R. and Kates, R. (1980) 'Handling hazards', *Environment*, **20** (7), in Dowie, J. and Lefrere, P. (eds), *Risk and Chance, Selected Readings*, Milton Keynes: Open University Press.
Foucault, M. (1979), *Discipline and Punish*, Harmondsworth: Penguin.
Foucault, M. (1980), *Power/Knowledge*, Brighton: Harvester.
Freymann, J.G. (1975), 'Medicine's great schism: prevention vs cure: an historical interpretation', *Medical Care*, **13** (7): 525–36.
Gough, I. and Doyal, L. (1991), *A Theory of Human Need*, Basingstoke: Macmillan.
Graham, H. (1979), 'Prevention and health: every mother's business, a comment on child health policies in the 1970s', in Harris, C.C. (ed.) in association with Anderson, M. et al, *The Sociology of the Family: new directions for Britain*, Sociological Review Monograph 28, Keele: University of Keele.
Grimshaw, P., Harvey, L. and Pease, K. (1988), *Crime Prevention Delivery: the work of police crime prevention officers*, Manchester: University of Manchester, Department of Social Policy and Social Work.
Hardiker, P., Exton, K. and Barker, M. (1989), *Policies and Practices in Preventive Child Care, Report to the Department of Health, Supplement II, Additional Literature Reviews: Prevention in Health Care, Crime Prevention*, Leicester: University of Leicester School of Social Work.

Hardiker, P., Exton, K. and Barker, M. (1991), 'The social policy contexts of prevention in child care', *British Journal of Social Work*, **21** (4): 341–59.
Hayek, F.A. von (1946), *The Road to Serfdom*, London: Routledge.
Heal, P. (1983), 'The police, the public and the prevention of crime', *Howard Journal of Penology and Crime Prevention*, **22** (2): 91–100.
Health Education Council (HEC) (1984), *Coronary Heart Disease Prevention. Plans for Action*, London: Pitman/HEC.
Herriger, N. (1986), *Präventives Handeln und Soziale Praxis. Konzepte zur Verhütung abweichenden Verhaltens bei Jugendlichen und Kindern*, Weinheim: Juventa Verlag.
Himmelstrand, U. (1960), *Social Pressures, Attitudes and Democratic Processes*, Stockholm: Almqvist and Wiksell.
Holman, R. (1988), *Putting Families First. Prevention and child care*, Basingstoke: Macmillan.
Holtzman, N.A. (1979), 'Prevention: rhetoric and reality', *International Journal of Health Services*, **9** (1): 25–39.
Hope, T. and Murphy, D.J.I. (1983), 'Problems of implementing crime prevention: the experience of a demonstration project', *Howard Journal of Penology and Crime Prevention*, **22** (1): 38–50.
Karmaus, W. (1981), 'Präventive Strategien und Gesundheitsverhalten', *Das Argument*, Argument Sonderband, **64**.
Klages, H. (1982), 'Prävention als Sozialutopie', in Herder-Dorneich, P. and Schuller, A. (eds), *Vorsorge zwischen Versorgungsstaat und Selbstbestimmung*, Stuttgart: Kohlhammer.
Knowles, C. (1991), 'Afro caribbeans and schizophrenia: how does psychiatry deal with issues of race, culture and ethnicity?', *Journal of Social Policy*, **20** (2): 173–90.
Kreuter, H. and Geiger, A. (eds) (1986), *Prävention – Vorbeugende Maßnahmen im Gesundheitswesen*, Köln: Deutscher Instituts-Verlag.
Labisch, A. (1982), 'Entwicklungslinien des öffentlichen Gesundheitsdienstes in Deutschland. Vorüberlegungen zur historischen Soziologie öffentlicher Gesundheitsvorsorge', *Öffentliches Gesundheitswesen*, **44**: 745–61.
Leonardis, O. de and Mauri, D. (1983), 'Der Prävention vorbeugen', in Wambach, M.M. (ed.), *Der Mensch als Risiko, Zur Logik von Prävention und Früherkennung*, Frankfurt: Suhrkamp.
Lewis, J. (1986), *What Price Community Medicine?* Brighton: Harvester Wheatsheaf.
Love, S.B. (1987), *The Economics of Crime Prevention*, Huntingdon: Cambridgeshire Police.
McGrath, M. (1988), 'Inter-agency collaboration in Wales?, *Social Policy and Administration*, **22** (1): 53–67,
MacIntyre, S. (1976), 'To have or to have not – promotion and prevention of childbirth in gynaecological work', in Stacey, M. (ed.), *The Sociology of the National Health Service*, Sociological Review Monograph 22, Keele: University of Keele.
Manning, N. (1987), 'What is a social problem?', in Loney, M. (ed.), *The State or the Market. Politics and welfare in contemporary Britain*, London: Sage/Open UP.
Moroney, M.J. (1951), *Facts from Figures*, Harmondsworth: Penguin.

Morris, J.N. (1975), *Uses of Epidemiology*, 3rd edn, Edinburgh: Churchill Livingstone.
Muir Gray, J.A.L. (1979), *Man Against Disease. Preventive medicine*, Oxford: Oxford UP.
Navarro, V. (1976), *Medicine under Capitalism*, New York: Prodist.
Newton, J. (1988), *Preventing Mental Illness*, London: Routledge.
Nocon, A. (1989), 'Ignorance and joint planning', *Social Policy and Administration*, **23** (1): 31–47.
Parton, N. (1985), *The Politics of Child Abuse*, Basingstoke: Macmillan.
Patton, C. (1990), *Inventing AIDS*, London: Routledge.
Pill, R. and Stott, N.C.H. (1982), 'Concepts of disease causation and responsibility: some preliminary data from a sample of working class mothers', *Social Science and Medicine*, **16** (1): 43–52.
Porter, D. and Porter, R. (1988), 'The enforcement of health: the British debate', in Fox, D.M., and Fee, E. (eds), *AIDS. The Burdens of History*, Berkeley: University of California Press.
Rapoport, L.R. (1961), 'The concept of prevention in social work', *Social Work*, **6** (1): 3–12.
Rawls, J. (1972), *A Theory of Justice*, Oxford: Oxford University Press.
Reiser, S.J. (1980), 'The emergence of the concept of screening for disease', *World Health Forum*, **1** (1,2): 99–104.
Renaud, M. (1975), 'On the structural constraints to state intervention in health', *International Journal of Health Services*, **5** (4): 559–71.
Richardson, A. (1983), *Participation*, London: Routledge and Kegan Paul.
Rodmell, S. and Watt, A. (eds) (1986), *The Politics of Health Education. Raising the issues*, London: Routledge and Kegan Paul.
Russell, L.D. (1986), *Is Prevention Better Than Cure?* Washington: Brookings Institute.
Sachße, C. and Tennstedt, F. (eds) (1986), *Soziale Sicherheit und Soziale Disziplinierung. Beiträge zu einer historischen Theorie der Sozialpolitik*, Frankfurt: Suhrkamp.
Schülein, J.A. (1983), 'Gesellschaftliche Entwicklung und Prävention', in Wambach, M.M. (ed.), *Der Mensch als Risiko. Zur Logik von Prävention und Früherkennung*, Frankfurt: Suhrkamp.
Schuller, A. (1982), 'Prävention als Sozialtechnologie', in Herder-Dorneich, P. and Schuller, A. (eds), *Vorsorge zwischen Versorgungsstaat und Selbstbestimmung*, Stuttgart: Kohlhammer.
Schwartz, F.W. (1982), 'Ziele und Strategien "präventiver Medizin"', in Herder-Dorneich, P. and Schuller, A. (eds), *Vorsorge zwischen Versorgungsstaat und Selbstbestimmung*, Stuttgart: Kohlhammer.
Smith, A. (1968), *The Science of Social Medicine*, London: Staples.
Squires, P. (1990), *Anti-Social Policy: Welfare Ideology and the Disciplinary State*, Hemel Hempstead: Harvester Wheatsheaf.
Stanworth, M. (ed.) (1987), *Reproductive Technologies: Gender, Motherhood and Medicine*, Oxford: Polity.
Stone, D. (1989), 'Upside down prevention', *Health Service Journal*, 20 July, 890–91.

Stone, D.A. (1986), 'The resistible rise of preventive medicine', *Journal of Health Politics, Policy and Law*, **11** (4): 671–96.
Susser, M. (1973), *Causal Thinking in the Health Sciences, Concepts and strategies of epidemiology*, New York: Oxford UP.
Swaan, A. de (1990), 'Expansion and limitation of the medical regime', in Swaan, A. de, *The Management of Normality, Critical essays in health and welfare*, London: Routledge.
Tawney, R.H. (1938), *Equality*, London: Allen and Unwin.
Taylor, R.C.R. (1982), 'The politics of prevention', *Social Policy*, Summer, 32–41.
Tesh, S. (1982), 'Political ideology and public health in the nineteenth century', *International Journal of Health Services*, **12** (2): 321–42.
Trojan, A., Soden, K. Von and Wallenczus, K. (1990), 'Selbsthilfeforschung und AIDS. Einige Konzepte, Probleme und Forschungsfragen', in Rosenbrock, R. and Salmen, A. (eds), *AIDS-Prävention*, Berlin: Rainer Bohn Verlag.
Tudor Hart, J. (1971), 'The inverse care law', *The Lancet*, 27 February, 405–12.
Tuomilehto, J. and Puska, P. (1987), 'The changing role and legitimate boundaries of epidemiology: community-based prevention programmes', *Social Science and Medicine*, **25** (6): 589–98.
Turner, B. (1982), *The Body and Society*, Oxford: Blackwell.
Vobruba, G. (1983), 'Prävention durch Selbstkontrolle', in Wambach, M.M. (ed.), *Der Mensch als Risiko, Zur Logik von Prävention und Früherkennung*, Frankfurt: Suhrkamp.
Waitzkin, H. (1989), 'Social structures of medical oppression: a Marxist view', in Brown, P. (ed.), *Perspectives in Medical Sociology*, Belmont, CA: Wadsworth.
Weeks, J. (1989), 'AIDS: the intellectual agenda', in Aggleton, P., Hart, G. and Davies, P. (eds), *AIDS: Social Representations, Social Practices*, Lewes: Falmer.
Wikler, D. (1978), 'Persuasion and coercion for health', *Milbank Memorial Fund Quarterly*, **56** (3), in Dowie, J. and Lefrere, P. (eds), (1980) *Risk and Chance, Selected Readings*, Milton Keynes: Open University Press.
Wilkins, L.T. (1985), 'The politics of prediction', in Farrington, D.P. and Tarling, R. (eds), *Prediction in Criminology*, Albany: State University of New York Press.
Williamson, J.D. and Danaher, K. (1978), *Self Care in Health*, London: Croom Helm.
Yarrow, A. (1986), *Politics, Society and Preventive Medicine. A Review*, London: Nuffield Provincial Hospitals Trust.

5 Health, harm or happy families? Knowledges of incest in twentieth century parliamentary debates

Vikki Bell

In the stark form of 'thou shalt not' the incest prohibition is enshrined in both English and Scots criminal law.[1] At first glance, this legislation simply reflects and reproduces a general and unquestioned belief that incest is fundamentally and unquestionably wrong. However, it is my contention that the debates which have addressed the criminalization of incest this century have not only relied upon the basic assertion that incest is wrong, but have drawn more and more upon disciplinary knowledges of incest. That is, incest has become an issue to which several knowledges are addressed such that the debates around it draw upon several different justifications for criminalizing it. Each knowledge, moreover, constructs incest as a particular kind of wrong, placing particular demands on the terms of the legislation. The result is that during the debates the object under discussion – 'incest' – shifts its meaning and its terms of reference according to the knowledge by which its 'wrong' is being asserted. Emerging from this battle of knowledges is the legal category 'incest', which, whilst dependent upon the knowledges that have shaped it, is not reliant upon any one of those knowledges to define its parameters: criminal law presents a further Truth of incest.

The parliamentary debates I consider in this chapter are from two different periods and deal with English and Scots law respectively. They are debates which deal solely with the issue of incest. The first set of debates are those that took place around the criminalization of incest in English law at the beginning of this century. Incest had not previously been a crime in English law.[2] It had been dealt with by the Church up until 1857, when the removal of power from the Church had left incest neither an ecclesiastical nor a criminal offence[3] (Wolfram, 1983). There were several unsuccessful attempts to pass this Bill before 1908 (in 1899, 1903 and 1907). The debates I have studied are those when the Bill was discussed at length.

They are those of 1903, when the Bill was unsuccessful, and of 1908, when it was passed. The second set of debates are those that took place in the 1980s and dealt with the revision of Scots law. In Scotland, incest has been criminal since 1567 but this Act was not updated until the 1986 Incest and Related Offences Act. With the parliamentary debates of the 1980s, I have also included the Scottish Law Commission's work (1980 Memorandum and 1981 Report) from which the Bill was taken.

Traditionally, incest has been declared 'against God's will' or regarded as a moral offence in a way which depicts incest as an act which transgresses some code of behaviour set down by an higher order, either God or the more abstract 'moral society'. Statements which depict incest as 'simply wrong' in this way do appear in the parliamentary debates in both periods. More interesting, however, is the way in which speakers speak about incest as a wrong, with reference to what one might term 'knowledges'. They are knowledges in the sense that there is a body of writing and a set of experts which accompany the assertions.

My study of parliamentary discussions of incest takes place within a framework derived from the work of Michel Foucault. I am drawing upon a general Foucauldian understanding of discourse as a site of struggle, a web of power/knowledge networks. There is room for a more detailed utilization of Foucault because the debates I am studying here are about sex, a subject which also attracted Foucault's scholarly attention. In *The History of Sexuality: An Introduction* (1981) Foucault argues that there has been an explosion of discourses concerned with sex: 'talking sex' might pass as the emblem of modern society. His concern was with the expansion of exactly the sort of knowledges of sexuality that I make the focus of attention here.

The knowledges I identify in the debates are ones which might be thought to form part of Foucault's 'deployment of sexuality', the movement which has surrounded the family and its 'sexuality' with 'experts' who watch out for and interpret signs of sexuality (and in doing so construct the sexuality of which they speak). In a sense then, the paper illustrates how incest has become a problem not only breaking moral and religious codes, but a problem requiring study. The justifications for criminalizing incest that take place within the parliamentary debates construct incest as a problem that can be measured and considered by rational, scientific methods.

In short, this paper is a study of the 'problematisation' of incest. Foucault explained this concept as follows:

> Problematisation doesn't mean representation of a pre-existing object, nor the creation by discourse of an object that doesn't exist. It is the totality of discursive or non-discursive practices that introduces something into the play of true or false and constitutes it as an object of thought (whether in the form of moral reflection, scientific knowledge, political analysis, etc.) (Foucault, in Kritzman, 1988:257).

The debates reveal how incest has become an 'object of thought' within several different discourses – medical, social welfare, feminist and so on – that have different understandings of what incest is 'about' and that have produced different sorts of knowledges of incest. When the different knowledges meet, as they do in these debates, the very parameters of what constitutes the 'truth' about incest are continually attacked and reconstituted.

In considering these debates, I do not want to get drawn into an analysis of their timing or causation.[4] Rather, the focus is on *how* incest is spoken about, and how the speakers use different knowledges to construct incest as a problem. In particular, I highlight how the debates construct incest as a specific crime. That is, how the different knowledges suggest that there is a specific wrong of incest such that it cannot be dealt with by other legislation (e.g. rape and sexual offence laws) but requires separate legislation. In doing so I am making an argument about the process of law creation.

Carol Smart (1984, 1989) has argued that that the law itself is a form of knowledge which has the power to disqualify other accounts of social reality. Smart considers how this happens in the rape trial, where the law 'consistently fails to "understand" accounts of rape which do not fit with the [law's] narrowly constructed definition' (1989:26). Whilst the woman survivor may be willing to tell 'the whole truth' of the event as she experienced it, her experience is only selectively heard through the parameters of what the law considers relevant. The law has the power to decide what it is necessary to know before a decision can be made about what 'really happened'.

In my analysis I take a step 'backwards' from the court room situation in order to consider the process of law creation, a process in which law does not yet have a knowledge from which it can 'disqualify' other knowledges, but is, instead, the site at which various knowledges, each presenting its 'truth' of incest, meet and map out a space for incest as a specific crime. I have divided the knowledges articulated in these debates into three sections. First I consider the problematization of incest as a problem of inbreeding; secondly, I look at those who see incest as wrong because it causes harm to a living individual and thirdly, those that see incest as a threat to the institution of the family. In tracing the articulation of these different knowledges in the debates, I shall not be assessing the relative truth of the knowledges, but indicating how they produce an attendant image of what incest is 'about'. These images will conflict as well as coincide. Ultimately, this discursive battle of knowledges is silenced and denied when the legislation is presented and understood as a coherent response to a singular problem. The paper is therefore an investigation of how the law's Truth is constructed at the level of the statute.

Incest as a problem of health: the inbreeding argument

One of the ways in which incest is presented in the debates is as a problem of health. According to this knowledge, what is wrong with incest is the potentially deleterious effects of inbreeding. The health in question is that of the potential off-spring and, on occasion, this becomes the health of the nation. Although this problematization was not voiced in 1903, in 1908 there were references to the risks of incest by four separate speakers. Here are two of those four:

> Mr. MacLean of Bath said ... he had known of instances producing no less than three or four children of weak intellect, idiots and imbeciles. The cases were of the most grave kind (26.6.1908, Commons).

> In a certain number of crimes of a similar character it might be argued that it would be desirable not to take steps with regard to them, because they affect nobody but those immediately concerned; but your Lordships will see that that is not the case with regard to crimes of this character, and that there are, as a result of intercourse between the various people mentioned in the Bill, off-spring on whom the punishment chiefly falls (The Lord Steward, Earl Beauchamp, 2.12.1908, Lords).[5]

In the 1980s, the inbreeding argument was still important and clearly articulated in the debates on the revision of the Scottish legislation. In the Scottish Law Commission's Memorandum (1980) and Report (1981) 'genetic reasons for prohibiting incest' are given as one of the four main arguments for retaining the crime of incest.[6] The Memorandum states:

> intercourse between certain persons should be prohibited because the offspring are more liable to exhibit physical and mental abnormalities (1980:22).

The studies upon which the Commission draws are ones which have observed and 'measured' children born to related parents, seeking to find quantifiable differences in terms of health but also in terms of mental ability and general 'intelligence' (the studies include Adams and Neal 1967; Seemanova 1971). These researchers are constructed as the experts on incest, the ones who have produced scientific 'truths' of incest. It seems that the inbreeding argument is still important in the creation of the 1986 Act. This is underlined when the Commission consider whether the crime of incest need remain a specific offence in criminal law, or whether other areas of law, such as rape or sexual offences would adequately deal with the wrong of incest. They state:

> In our view, to have no such provision would be to take less than proper account of genetic considerations (1980:59).

Several commentators have argued that on its own the inbreeding argument is insufficient justification for criminalizing incest (Mason 1981; Wasoff 1980). Amongst several objections to the inbreeding argument as the basis of law, is the argument that legislation based on such an objection to incest would logically have to disregard cases where there was no danger of conception, e.g. because one or both parties is unable to conceive for medical reasons, for reasons of health or age (too young or too old). This limits the criminality of the act to specific groups (the fertile) and specific times (during fertile years, possibly even fertile days). It further limits the criminality to the consequences of the act, suggesting that if no pregnancy follows, the act itself was not wrong. It suggests there is nothing intrinsically wrong with incestuous intercourse, but the possible consequences are such that it should be criminalized. Yet despite the arguments that have questioned the inbreeding argument, it is repeatedly rehearsed when incest is discussed.

My question with regard to the inbreeding arguments is not how important was the inbreeding argument in the passage of the Bill? Nor do I ask: is it relevant or even accurate (in the past or now)? Instead, I ask: how does the inbreeding argument function in the parliamentary debates? How does it inform the meaning of the term 'incest', and how does it relate to the process of mapping out a place for incest as a specific crime? The presence of this knowledge in the debates brings with it some important correlatives that construct a particular image of incest. There are four important points:

- i) It constructs and relies upon an image of incest as sexual intercourse between a man and a post-menses woman, and, therefore, as heterosexual and between adults (in the sense of being able to conceive). When the 1981 Report considers including 'homosexual' offences, it is the genetic argument that is used to argue that they should not be included.
- ii) It constructs incest as only between people related by blood ('full' or 'half' blood). Step-father/step-daughter incest, for example, is not included in its image of incest.
- iii) The 'victim' of incest is depicted as the potential offspring.
- iv) Incest becomes a specific wrong. Thus incest is wrong for reasons that other behaviours are not. The wrong of incest is something other than the wrong of rape, for example, and therefore needs to be separated from rape. It is in this sense that the place of incest is being discursively 'mapped' out. The inbreeding problematization of incest gives it a specificity that disallows its collapse into other categories of offence.

Despite the importance given to the problematization of incest as a problem of health-risk for the potential offspring, in the Law Commissions work, in the parliamentary debates of 1985 and 1986 it is hardly articulated, possibly because the debates tend to centre on how to include step-children in the Bill (where genetic considerations would be out of place) with blood relatives presented as unproblematic. There is, however, an interesting exchange in the Standing Committee in which the continuing importance of the genetic argument in the justification for criminalizing incest is highlighted. When the first speaker constructs incest as a problem of non-consent, this understanding of the wrong of incest is challenged. The genetic argument is then offered (as well as linking incest to the cohesion of the family), acting as the guarantee of the specificity of incest:

> Mr. Fairbairn: ... It seems that it is only when there is abuse rather than consent that one should generally expect a prosecution to arise ...
>
> Mr. Maxton: ... In the latter part of his speech, the hon. and learned Member ... began to talk about the law having to deal with abuse in terms of incest. He began to sound as if he were saying that incest is all right provided both sides consent, and that we should prosecute only when there is abuse. That is not the view of the law ...
>
> Mr. Fairbairn: ... If the law is to be sensible, it will not prosecute wrongs because they are nameless wrongs, but wrongs which do harm. The prosecution must always bear in mind the wrong that one is trying to strike at, which either is a threat to the cohesion of the family or would lead to genetic deterioration [sic] ... (11.6.1986:13–14).

Thus, when challenged on his understanding of incest as wrong because it involves abuse, the Member reverts or defers to the genetic argument. The inadequacy of this problematization is pointed out in the reply.

> Mr. Maxton: ... I give the example of a man of forty who marries early and has a daughter when he is nineteen. His wife dies and he is a widower: he has a daughter now of twenty-one. If he sleeps with her for the first time and there is consent between them, does the hon. and learned gentleman say that that is legitimate because he may have had a vasectomy and therefore is not capable of procreation and there is no danger to the species? (11.6.1986:13–14)?

Moreover, this exchange illustrates the way that different constructions of the problem of incest shift the meaning of incest and the image of the incest situation that accompanies it.

Reading these debates through Foucault, one might see the references to offspring as part of the movement he has described whereby

> governments perceived that they were dealing not simply with subjects or even with a 'people', but with a 'population' with its specific phenomena and its

> peculiar variables: birth and death rates, life expectancy, fertility, state of health, frequency of illnesses, patterns of diet and habitation (Foucault 1981:25).

At the heart of this economic and political problem of population, Foucault continues, was sex. In order to 'know' and keep in touch with the population, it became necessary to analyse

> the birthrate, the age of marriage, the legitimate and illegitimate births, the precocity and frequency of sexual relations, the ways of making them fertile or sterile, the effects of unmarried life or of the prohibitions ... (1981:24–5)

The references to the effects of inbreeding, therefore, could be connected to the operations of a style of governing, one that watched over the health of the nation through the study of its life and its sex. Whilst these debates do not give us any real evidence that this was new, it does seem that the inbreeding argument has all the aspects that Foucault associates with this new form of government with its use of what Foucault termed 'bio-power' or, alternatively, 'pastoral power' (1981, 1982). This form of power is bipolar. At the one end, there are techniques directed toward the population as a whole; at the other, there are disciplinary techniques, directed toward the individual body.

The studies of inbreeding, both those that were around at the time of the English debates and the more recent studies quoted by the Scottish law Commission, display these disciplinary techniques. Nineteenth century studies looked at close breeding communities, measuring their bodies and recording the results in a scientific mode, seeking any differences that could be attributed to the inbreeding.

> At Boyndie, in Banffshire, and Rathen, in Aberdeenshire, the fishermen are a very closely bred community, the former somewhat shorter and lighter than the landsmen, and their heads are not quite so big (Huth 1875:150, reporting a study by Beddoe).

The measuring and observing did not always support the inbreeding hypothesis, however:

> [In Burnmouth and Ross, Berwickshire] there was no case of a lame, deformed, blind, dumb or paralytic person to be heard of; and in the school, which was twice visited, and where nearly all the children were assembled, no strumous sores were found, nor were any of the children puny, pale or languid, but on the contrary, merry and active (Huth 1875:148, reporting a study by Mitchell).

Similarly, the later studies on which the debates of the 1980s draw, investigate the children of incestuous unions, using the more 'sophisticated' measures such as IQ tests, but with the same hypothesis at work and with the same disciplinary gaze that seeks out quantifiable differences to report and present as knowledge, as scientific truths.

Incest as a problem of abuse: the protectionist, psychological and feminist arguments

In this section I consider those arguments that see incest as wrong because of an immediate abuse of one or more of the parties involved. It is a harm against a living individual or individuals.[8] This understanding of what incest is 'about' is one that clearly conflicts with the inbreeding argument, and indeed, that would render the inbreeding argument irrelevant, since whether or not there is a danger to a further individual (the offspring) in no way changes the fact of the immediate harm done to the individual concerned. This would be the wrong that feminists see in incest, although it is not the case that whenever incest is constructed as wrong in this way, it reflects a feminist knowledge of incest. I have identified three ways in which incest is spoken about as an immediate harm (these are not necessarily mutually exclusive): in arguments that equate incest with child abuse; in arguments that stress the psychological effects of incest, and in feminist arguments that stress the gender differential in incest. I shall deal with these in turn.

Child abuse

The 1903 Bill, in its original form, constructed an image of incest as older male/younger female insofar as the criminalized relations for a man were daughter, grand-daughter and sister, and for a woman they were father, grand-father and brother.[9]

In the 1908 debates incest was also constructed as an attack upon children by an adult man. There was, for example, an amendment that proposed to remove guardianship or authority from a man convicted of incest against a female of whom he was legally guardian (Sir Samuel Evans, 26.6.1908, Commons). The expertise and the knowledge of the National Society for the Prevention of Cruelty to Children were brought into the debate in order to illustrate that cases of incest do occur (The Lord Bishop of St Albans, 2.12.1908, Commons). These agents of a Foucauldian 'pastoral power' not only provided evidence that incest did occur but framed it within their particular concern for children.

A telegram from the Lord Chief Justice read out in the 1908 Lords debates, also depicted incest as an assault by a man upon a younger female:

> You can state that I support the Bill. I have received and sent to the Home Secretary presentments of grand juries pointing out the urgent necessity for an

amendment of the law, in consequence of the frequency of assaults by fathers on their daughters (read out by the Lord Bishop of St Albans, 2.12.1908).

Such a construction of incest is not always used in favour of the Bill, however, since it is also argued that an Incest Bill was not necessary because most cases could be dealt with under the Cruelty to Children Acts or under the Criminal Law Amendment Act (The Earl of Crew, 2.12.1908, Lords).

The notion that incest is a form of child abuse from which children need protection is also present in the 1980s debates. Speaking of raising the age to which step-children are included as a 'related offence' from 16 to 21, it is said:

> This amendment is to add to the protection which the young person, the step-child, would have in the family (Lord Morton of Shuna, 28.1.1986, Lords).

The psychological impact

In the 1980s debates, however, a possibly more dominant problematization of incest is that which draws on psychological knowledge. In the Scottish Law Commission's (1980) Memorandum, there is a chapter entitled 'The Psychological Effects of Incest', and this understanding of the wrong of incest is given much space in that document. Here the wrong is represented by the psychological problems incest has for the female survivor. The studies which report the psychological effects of incestuous abuse, continue a disciplinary gaze, and often one which positions the abuse as a disruption of 'normal development'. In so doing, there are assumptions being made about what counts as 'normal development', particularly sexual development. For example, the Memorandum states that the female survivor of incest is endangered by the 'premature development of sexuality without adequate means of coping with the sexual tension' (1980:37). Drawing on further scientific studies, it suggests she is then prone to seek outlets for her inner turmoil by way of anti-social behaviour and drug use.[10]

In the psychological problematization, moreover, incest may be seen as a problem that will lead to the female survivor becoming a 'problem' for society (she will become 'anti-social'). In this way, interestingly, the psychological knowledge of incest constructs it as a threat to future generations, in the same way that the inbreeding argument does. Drawing on a study of incest as a 'causative factor' in anti-social behaviour,[11] it is argued in the Memorandum that

> the result could well be a second generation of inadequate persons who will produce subsequent generations of neglected children unless we develop tools for prevention, detection and treatment of these families and their children (1980:36).

Moreover, the psychological knowledge often brings with it implicit theories about the causes of incest, further depictions of what incest is 'about'. In the Memorandum, and in the later Report (1981), one way in which incest is described is as a problem of family functioning whereby certain families are set up as dysfunctional. The Scottish Law Commission's (1980) Memorandum quotes at length from a study by Maisch (1972). This work saw incest as a result of 'family disorganization' which had preceded it. Each member of the family, but particularly the parents, is depicted as contributing in some way to the incest. The disorganization is described with reference to:

> the negative influence of the husband on the shaping of the marriage and the family in the first place, a similar negative influence on the part of the wife; violence and irascibility often associated with drinking on the part of the father; and promiscuity, an unsettled way of life, drinking and drug-taking on the part of the wife together with her physical illness often leading to absence from the home. The daughters of these parents evinced symptoms of disturbed personality development either in the form of psychosomatic symptoms, dissociality (as for e.g. truancy, running away from home, frequent lying, undesirable sexual relations) or neuroses and other behavioural disturbances (for e.g. anxiety symptoms such as fear of death, claustrophobia and suicidal tendencies) or depression (1980:32–3).

The feminist critique of such psychological assertions has highlighted the way in which the notion that the particular families in which incest occurs are dysfunctional provides a way of taking the responsibility off the abuser, as well as very often blaming the mother as in the above description, for example, listing the mother's illness and absence as evidence of her contribution to 'disorganization'. There is a psychologization of incest that depicts the family as in some way 'abnormal'. Simultaneously, there is a normalization strategy within this type of knowledge, since there is an implicit model of organization to which this disorganization is contrasted. But it is the normal Family itself that feminist knowledge places at the centre of the problem.

The feminist perspective

The feminist knowledge of incest is not incompatible with aspects of the previous two understandings of the wrong of incest. However, some their aspects are quite different, especially where the implicit causes in the psychological knowledges quoted by the Law Commission are concerned. I shall not provide a detailed survey of feminist analyses of incest here, but suffice to say that feminists connect incest with the more general power structure between men and women, and analyse it with reference to the construction of 'normal' male sexuality and 'normal' family structure.

In the early debates, a feminist understanding of incest is not explicitly articulated. In the 1980s debates, however, there are some comments that present a feminist knowledge. An amendment to raise the age at which a step-child is included within the crime of incest is backed up with reference to an argument that challenges the concept of consent, which feminists would welcome (although the comparison this speaker makes with unrelated rape would need to be challenged as well):

> it is an open question whether one can usefully talk in terms of a daughter's or son's consent, particularly if one takes into account the fact that the young girl in an incest situation is subject to a completely different set of conditions regarding defence, tolerance and participation from the child or maturing girl who meets a completely unrelated adult aggressor (S.L.C. 1981, quoted by Lord Morton of Shuna, 28.1.1981, Lords).

There is only one speaker, a woman, who constructs her argument from an unmistakeably feminist knowledge of incest. She tells the House:

> I have been forced to be interested because of the increasing and in many ways, welcome discussion about incest, which until fairly recently has been swept under the carpet. We should all welcome the fact that now both men and women are beginning to discuss these problems, which are particularly difficult for the girl or young women involved in what amounts to violence against her by a member of her family ...
>
> Many women of all ages have told me of the pressures on them from their fathers, stepfathers or in some cases, their elder brothers to develop a relationship. It can start at an early age when the young girl does not know when it is happening, is frightened, and knows something is wrong but does not know what (Ms Richardson, 4.7.1986, Commons).

Tracing the construction of incest as a harm against a living individual, therefore, leads one into a consideration of the child protection, the psychological and the feminist arguments about what is wrong with incest. The 1903 Bill did seem to reflect an understanding of incest as older male/ younger female, but the debates do not reveal a feminist knowledge of incest. In the early debates, the construction of incest as a harm against a living individual is represented on this parliamentary level by the protectionist argument. In the Scottish Law Commission, the psychological arguments are accorded the most space of these three constructions of incest as an immediate harm. It is not until the 1986 Commons debates that a feminist knowledge seems to be explicitly drawn upon and incest is constructed as violence against young women and children within the family (Ms Richardson, 4.7.1986, Commons). In a broad sense, these understandings of the wrong of incest are compatible. But they do differ in the image they construct. The victim of incest in the psychological studies, for example, is sometimes the family as a whole, as opposed to one individual. The protection arguments stress the age difference, the fact that it is the category 'children' who are at risk. This would not be compatible with a feminist perspective which considers the fact that the abused is very often 'a child' is as important as the balance of power relations in the situation. In feminist analyses the wrong of incest is constructed as on a par with rape, as sexual abuse related to the gendered power dynamics of a male dominated society that depicts women and children as possessions of the male (Herman 1981; Ward 1984).

Happy families: incest as a problem of the family

Another major way in which incest is constructed as a problem is the construction of incest as a problem of the family. I have already mentioned the way in which incest is seen as the result of a disorganized family system in some of the psychological studies. In this section, however, I want to look at the way incest is constructed as wrong because it threatens the *institution* of the family.

In the early debates, although the family was spoken about as under threat, it was the criminalization of incest that was set up as the danger to the family and not the practice of incest itself.

> Poor people were forced to live together, men and women often in the same room, and it was easy enough for anybody to make accusations which could not be disproved against parties who were totally innocent. They would make family life impossible if they passed such legislation. They would make it impossible for members of a family to live at home without running the risk of terrible accusations being brought against them (Mr Lupton, 26.6.1908, Commons; see also Mr Staveley-Hill, 26.6.1908, Commons).

In the 1980s debates, by contrast, the protection of the Family is a recurrent understanding of the purpose of the crime of incest. In the Memorandum the authors note that 'there is a core of theory associating the prohibition [of incest] with the need to preserve the family'. Echoing a sociological functionalist line of argument[12] they state:

> it [the family] is the fundamental unit of all larger social groups and the principal means of preparing the child to participate as a mature adult in the life of his community (1980:3).

In order to preserve the institution of the family, the Memorandum depicts the criminalization of incest as 'controlling the potentially disruptive cross-sex attractions and rivalries engendered within the family' (1980:3).

It is interesting to note the way in which the assumptions of this knowledge are overlaid upon the law of incest. In these arguments around preserving the family, it is cross-sex attractions that are held to be the danger. This may be because of the influence of Freud on functionalism through the work of Talcott Parsons, since Freud's Oedipus complex was based upon the assumption that unconscious attractions are always to the opposite sexed parent. The Parsonian knowledge sets up the potential attractions in the family as only cross-sex, and thereby provides a justification for the fact that the crime of incest involves only male/female vaginal intercourse. This is also true of the inbreeding argument, that relies upon incest as male/female. These knowledges did not *cause* the crime to be so limited because they would be post hoc justifications. Nevertheless, they construct the image that 'fits' the crime.

Furthermore, the construction of incest as wrong because it threatens the institution of the family, is seen in the Report's argument that relationships of affinity can and should be removed from the crime of incest because, inter alia,

> there is less risk of harm to the solidarity of the family since in-laws rarely form part of the typical family household in contemporary Scottish society (1981:19).

In the Standing Committee too, the exclusion of relationships of affinity is presented as a response to the changing family structure in Scotland, with the implication that they no longer form a part of 'the Family' which the Bill seeks to preserve:

> the genetic argument is not there because the parties are not related by blood, ... there are other provisions to protect children, and ... *increasingly in Scotland there is not the same family structure as in times past* (Mr James Wallace, 11.6.1986:4, emphasis added).

The inclusion of adopted child/adoptive parent relationship as incest in the 1986 Act might also be interpreted as relying on a problematization of incest as a threat to the Family, because it disregards genetic links within it and places more emphasis on social links.[13] In the Standing Committee, the way in which the inclusion of adopted children is presented constructs this understanding of the wrong of incest:

> Admittedly there is no blood tie, but because of the other factors which society deemed important and *the reason why the law of incest should exist – principally the question of trust and the family bond* – it was thought that the adopted relationship was such that it should come within the crime of incest (Mr James Wallace 11.6.1986:4, emphasis added).

In this way, then, incest is constructed as a problem because of the wrong it causes the institution of the Family in its present form. This problematization maps out a space for incest as a specific crime because the law becomes understood as part of the incest prohibition, which in turn is constructed as the basis of the family structure on which society is built. The crime of incest has therefore to reflect the family structure that it wishes to protect (hence, for example, the relationships of affinity are not included because they are not considered part of that family structure).

Conclusions

In this paper I have discussed three of the most important ways in which incest has been constructed as a problem in these parliamentary debates: incest as a problem of health, incest as a problem of harm to a living individual (which I divided into three), and incest as a threat to the institution of the Family. These are three understandings of the wrong of incest that are recurrent in the debates. They are not, of course, the only ways in which incest is spoken of as a problem but they are the constructions that are built upon what might be defended as knowledges of incest in the sense that there is frequently a body of writing, studies and 'experts' that accompany them.

Behind the bald criminalization of incest found in criminal law therefore, there is a web of interweaving knowledges and ways of understanding incest as a problem. In a Foucauldian schema these knowledges belong to the deployment of sexuality in that they are part of the will to know about sex. They are comprised of studies that seek to find out about incestuous sex, to establish a knowledge about it.[14]

British law on incest has been built up from these several different knowledges of incest. None of these is the sole knowledge upon which the law is based. The differences between the various problematizations of incest in the debates are not, in fact, addressed as differences. Their coexistence in the debates, and the fundamentally different positions that they may represent or suggest is not acknowledged within the debates. Indeed, the coalition of several knowledges has probably been instrumental in the process of law creation, with the different knowledges building and contributing to the forming of a law which may not have been so created if only one knowledge were articulated. Thus where the inbreeding argument breaks down, the Family argument can be brought in to protect the place of incest in the criminal law. In this way the law is reliant upon other knowledges. Indeed, the law's position, in both English and Scots law, might be described as eclectic, taking aspects of various knowledges in building its position. In this, I am echoing Smart (1984, 1989), who has suggested that the law has to be understood as complex and contradictory, with little of the autonomy or coherence it purports to have.[15]

These knowledges are tied to power. The legal discourse on incest has been a meeting place for relations of power/knowledge. Disciplinary techniques of power have been operating in the formation of the knowledges that have then met at the site of law. But the analysis of these debates illustrates that the legal discourse is not impenetrable. The challenge of the feminist knowledge on incest can and has been represented in the process of law creation. It has provided another way of conceptualizing incest which has received space in the discussions. Yet the 'discursive battle' that took place between the various knowledges of incest established none as victor, but rather, produced a further Truth, a powerful truth that subsequent legal procedures take as their point of departure.

Notes

1 Incest is defined as penetration by the vagina of the penis. Consent is no defence. The relationships criminalized in English law are grandfather/granddaughter; father/daughter; brother/sister and mother/son. 'Sister' includes half-sister and brother' includes half-brother. In Scots law there are further relationships criminalized as incest: grandmother/grandson; aunt/nephew; uncle/niece; great grandmother/great grandson and great grandfather/great granddaughter. Father/daughter and Mother/son by adoption are also criminalized in Scots law. There is also a clause to criminalize step-parent/stepchild; and an offence of 'intercourse with a person whilst in a position of trust' contained within the Scots law of incest.

2 Except for the period 1650–1661 when the Puritan Commonwealth made incest punishable by death. Apparently, however, the 1650 Act was largely a 'dead letter' (Bailey and Blackburn 1979).

3 Although the 1857 Matrimonial Causes Act did specify 'incestuous adultery' as grounds for divorce.

4 Subjects which have been debated by scholars with regard to the English legislation. See Bailey and Blackburn (1979); Wolfram, (1983); Jeffreys (1985).

5 See also The Under-Secretary of State for the Home Department, Herbert Samuel, 26.6.1908, Commons and The Lord Chancellor, Lord Loreburn, 2.12.1908, Lords.

6 The others are: maintaining the solidarity of the family; prevention of psychological harm to family members and the opinion that society as a whole would wish incest to remain criminal.

7 This question is left unanswered.

8 This phrase is used to separate these constructions of the wrong of incest from those that see incest as wrong because of the effects upon future generations or upon an abstract institution such as the family. I am aware that the titles of my typology is slightly confusing because both of these latter constructions are also arguments that incest does harm.

9 Bill 51. Later, son was added to the female's list of criminalized relationships.
10 Here the SLC draw on the work of Benward and Densen-Gerber (1975).
11 Benward and Densen-Gerber (1975).
12 Parsons (1954).
13 The Law Commission had initially recommended against this inclusion (in the Memorandum), but were persuaded otherwise by the commentators on that Memorandum, and the recommendation for inclusion is made in the Report.
14 This point, whilst drawing on Foucault, also echoes criticism of him for implicitly mapping repressive 'old' modes of power onto law. Law does not operate in a fashion opposite to bio-power, but is itself the site of the operations of power/knowledge.
15 For further discussion of how this perspective differs from other feminist analyses of law see Naffine, N. (1990).

References

Bailey, V. and Blackburn, S. (1979), 'The Punishment of Incest Act 1908. A case study in law creation', *Criminal Law Review*, p 708.

Benward and Densen-Gerber (1975), 'Incest as a causative factor in anti-social behaviour: in exploratory study', *Contemporary Drug Problems*, **4**.

Foucault, M. (1981), *The History of Sexuality: An Introduction* (trans. Hurley, R.): Harmondsworth Penguin (this translation first published 1979, Allen Lane).

Herman, J. (1981), *Father Daughter Incest*, Harvard, Mass.: Harvard University Press.

Huth (1875), *The Marriage of Near Kin Considered With Respect to the Law of Nations, the Results of Experience and the Teachings of Biology*, London: Longmans, Green and Co.

Jeffreys, S. (1985), *The Spinster and Her Enemies: Feminism and Sexuality 1880–1930*, London: Pandora.

Kritzman, L. (ed.) (1988), *Politics, Power, Culture: Michel Foucault. Interviews and Other Writings 1977–1984*, New York: Routledge.

Maisch, H. (1972), *Incest*, New York: Stein and Day.

Naffine, N. (1990), *Law and the Sexes*, London: Allen and Unwin

Parsons, T. (1954), 'The incest taboo in relation to social structure and the socialisation of the child', *British Journal of Sociology*, **5**.

Scottish Law Commission (1980), *The Law of Incest in Scotland*, Memorandum No. 44.

Scottish Law Commission (1981), *The Law of Incest in Scotland Report on a Reference under section 3 (1)(e) of the Law Commissions Act 1965*, Cmnd 8422, Edinburgh: HMSO.

Smart, C. (1984), *The Ties That Bind: Law, Marriage and the Reproduction of Patriarchal Relations*, London: Routledge and Kegan.

Smart, C. (1989), *Feminism and The Power of Law*, London: Routledge.

Ward, E. (1984), *Father–Daughter Rape*, London: The Women's Press.

Wolfram, S. (1983), 'Eugenics and the Punishment of Incest Act 1908', *Criminal Law Review*, p 308.

6 The gaze of the counsellors: Discourses of intervention in marriage

David Clark and David Morgan

Concern about the state of marriage has grown during the twentieth century and it is now unusual for a month to pass without some press coverage highlighting public statements. the latest statistics on marriage and divorce or, indeed, some relevant piece of sociological research. Such a focus has, of course, continued alongside more popular examinations of the marital upsets of public figures, which, in the case of members of the royal family, take on something of the status of social dramas, as occasions to rehearse and redefine the boundaries between public and private and between freedom and obligation. Marriage is seen as being more risky than in the past and divorce rates as a potential source of public danger. The growing concern about the state and future of marriage has been contributed to and accompanied by the growth of a whole range of specialized institutions and professions (or near professions) with particular interests in, and expertise relating to, marital problems. In Britain, one of the leading agencies with a concern for marital problems has been the National Marriage Guidance Council which recently celebrated its fiftieth birthday and changed its name to 'Relate'.

We do not see the relationships between the growth of marital agencies and professionals in terms of a simple response to perceived social dangers and public concerns. Indeed, the notion of 'public concern' is always a highly problematic one, whether one is talking about marriage and divorce or about some other putative problem such as football hooliganism. Rather, we see the relationship between trends in marriage and the growth of marital agencies, such as Relate, as being mutually constitutive; for agencies not only respond to marital problems but they also sharpen and give definition to these very problems. In this chapter, we develop a more problematized understanding of the notion of 'marital problems'. We then go on to consider two possible constructions of these inter-connections between marital issues and marital agencies: the idea of 'medicalization' and the idea of the 'counselling gaze'. Both these possible constructions, we argue, take us some way into the analysis but both have their limitations and, in the concluding section, we discuss the possibilities of thinking in terms of a 'marital gaze'.

The social construction of marital problems

It will be remembered that C. Wright Mills chose marriage as an illustration of the contrast between private problems and public issues (Mills 1959), a division which itself might be seen as having more than an echo of Durkheim's distinction between individual acts of suicide and the suicide rate (Durkheim 1951). The pains and uncertainties which spouses experience when they feel that there is something 'wrong' with their marriage are to be distinguished from more collective concerns when divorce rates reach levels which come to be defined as problematic. Put another way, a divorce rate may be seen as being not only an aggregation of human unhappinesses but also a signifier of some underlying social malaise, indicative of some risk to the fabric of familial relations.

An illustration of the routine construction of divorce as a public issue may be seen in the following extract from a story from *The Guardian* dealing with Relate's annual review for 1989/90:

> The review shows that Britain has the second highest divorce rate in Europe – 12.6 per 1,000 marriages, after 12.7 in Denmark. All other countries are below nine per 1,000. It argues that divorce and separation cost £1.3 billion annually in legal aid, social security and housing, while Relate's work saves £40 million ... (*The Guardian* 17.10.90).

The fact that individuals, institutions and news publications routinely draw attention to the divorce rate and often, as is the case here, contrast this rate unfavourably with those in other countries, is itself a clear indication that divorce is being constructed as a 'public issue'. The shift in this story is away from individual unhappinesses and towards the social, indeed economic, costs of divorce in the country as a whole. Nevertheless, the link with individual problems is not entirely lost. Relate, as an organization committed to some considerable degree to individual and couple counselling, itself constitutes a link between the understanding of private problems and the construction of public issues. Indeed the story goes on to note that 'more than half of those who see Relate stay together' (ibid).

In dealing with the developing understanding of divorce as a social problem rather than a simple assemblage of individual problems, there are some tempting parallels to be drawn with well-established sociological understandings of deviance and the growth of 'moral panics'. Thus a shift might be made, in this kind of analysis, away from questions about the causation of increasing or high divorce rates and their possible regulation to an investigation of societal responses to divorce, the processes whereby they are constructed and amplified and the wider consequences of these understandings. In so far as an analysis of causes is proposed, it is in terms of the causes of these responses and their amplifications rather than of the initial 'problem' itself. (For one of the most elaborated examples of this mode of analysis see Hall et al 1978.) The history and analysis of moral panics concerning divorce remains to be written: the point to emphasize at this stage is that public concern about divorce is not simply a concern about individual unhappinesses. Furthermore, the term 'public concern' (all too readily used in this context) needs to be unpacked in order to identify the influential individuals and agencies that are involved in orchestrating and mobilizing this concern.

What is perhaps missing, in this suggested application of some version of 'labelling' theory to the study of divorce, is a reference to the other side of the coin, namely the individual 'personal problems'. To say that moral concerns about divorce rates are not simply reduceable to concerns about individual miseries is not to say that the latter are unimportant and nor is it to say that these issues are in some way off limits to sociological analysis. Indeed, we would argue that it is not possible to draw a straightforward line between divorce as an individual problem and divorce rates as a societal issue. The two are related and there is a flow between them which is more than a simple process of aggregation. At one level, couples who are undergoing a process of marital separation or divorce are not unaware of wider societal concerns and these, in some measure, may influence their understandings of their personal situation (Burgoyne and Clark 1984). Similarly, as we have seen, societal responses to divorce rates will normally include recommendations or support for more effective or more comprehensive systems of individual counselling or guidance. The analysis of the social construction of divorce rates as a societal issue therefore needs to proceed hand in hand with an analysis of individual understandings of divorce and marital problems.

That people have troubles, problems, difficulties, crises and upsets hardly needs to be stated. However, the extent to which and the ways in which varying and differing untoward circumstances are grouped together to constitute a set described as 'problems', is almost certainly subject to cultural variation. For something to be classed as a problem is not simply to identify it as a source of difficulty to the individual concerned but to place this identification in terms of that individual's historically and biographically situated projects. To bring together a series of life-events such as illnesses, bereavement, financial setbacks or brushes with the law as constituting instances of a broad class defined as 'problems' implies some readiness or capacity for abstraction that may not be found necessary in all cultures or all historical social situations.

Similar, if more focused, observations may be made about that class of troubles distinguished by the label 'marital problems'. It may be assumed that individual spouses have always had, potentially at least, some kind of grievance about their partners. Accounts in fiction as well as in 'fact', of nagging wives, brutal or unfeeling husbands and unfaithful husbands or wives are commonplace in the history of marriage. But these do not necessarily constitute 'marital problems' in the modern sense of the term and it is doubtful whether this term or some equivalent would have had much currency in the not too distant past. For example, consider this well-known description of a fictional boorish husband:

> He came home this morning at his usual hour of four, wakened me out of a sweet dream of something else, by tumbling over the tea-table, which he broke all to pieces; after his man and he had rolled about the room, like sick passengers in a storm, he comes flounce into bed, dead as a salmon into a fishmongers basket; his feet cold as ice, his breath hot as a furnace, and his hands and his face as greasy as a flannel night-cap. O matrimony! He tosses up the clothes with a barbarous swing over his shoulders, disorders the whole economy of my bed, leaves me half naked, and my whole night's comfort is the tuneful serenade of that wakeful nightingale, his nose! (Farquhar, *The Beaux-Stratagem*).

A disagreeable husband, no doubt, but would Mrs Sullen constitute this as a marital problem? For one thing, such complaints however familiar and however repeatable, focus upon particular individuals and their failings. Similarly, there have 'always' been complaints about the institution of marriage, about the unfavourable contrast with the supposed freedoms of single life and about the economic costs that it brings. But again these very widespread complaints do not appear to signify marital problems as the term might be used today. We may say that these more traditional complaints refer more to the institutional character of marriage – Mrs Sullens 'O matrimony!' – rather than to the relational aspects of marriage. In contrast, an example of a more relational understanding of marriage comes from Brannen and Collard's study of 'Marriages in Trouble':

> I think the most valuable thing I could say is that, having lived with two women and then married them, both changed after marriage, and maybe I changed as well, got sloppy, got lazy (Brannen and Collard 1982:69).

The commonly heard account, 'we just grew apart' may be a more straightforward reflection of a relational understanding and certainly has the hallmarks of a peculiarly 'modern' rationale for ending a marriage – displaying origins in psycho-therapeutic discourse.

Brannen and Collard's study, while subtitled 'the process of seeking help' is in fact a valuable illustration of the need not to take the definition of marital problems at face value. They are concerned with the processes whereby marital problems are defined and attributed, and the complexities whereby spouses or ex-spouses re-construct and re-interpret their pasts. They note that a high proportion of their sample did not initially understand their 'problems' as 'marital problems'; they were more likely to see them in individualistic terms. The realization of the existence of a problem and the reading of this problem as something to do with the marital relationship were two distinct stages in the 'marital careers' of the couples themselves. At a more societal level, Brannen and Collard are also concerned with the influential agencies and groups that are involved in the shaping and understanding of marital problems (ibid).

It is argued, therefore, that while individuals in any age may have been far from understanding or accepting marriage as a bed of roses, the construction of 'marital problems' may be taken as being historically specific. The term 'marital' suggests that the difficulties relate to features of the relationship itself rather than to any individual spouse, while the word 'problems' suggests that there are affinities between such difficulties as might be experienced or understood by different sets of spouses. While marital problems may, indeed will, be experienced in all their uniqueness and immediacy there is a sense in which the use of this term 'marital problems' implies that the difficulties and tensions are not, any longer, the property of any one particular couple.

'Marital problems', therefore, may be said to have a variety of features in common. In the first place, they may be said to be attached to the relationship itself. This, in its turn, may be seen as having a variety of dimensions. There is (as has already been noted) an emphasis upon the relational as opposed to the institutional features of marriage. Some well-established sociological accounts have argued for a general historical movement from the latter to the former; we prefer to see marriage as compounded of different mixes of the institutional and the relational (Lewis, Clark and Morgan 1992; Morgan 1990). Furthermore, as has already been argued, this emphasis upon the relationship implies a shift away from the identification of blameworthy partners within a marriage and a focus upon those features of a relationship that cannot be reduced to individual faults or failures. Put another way, individual pieces of behaviour (infidelity, alcoholism, sexual difficulties etc) may be seen as signifiers of relational problems. It is not, therefore, that individual behaviour is ignored in the construction of marital problems; rather, such behaviour is recast and others are persuaded to see it in a different light.

In the second place, 'marital problems' refer to a new construction of the ever fluctuating boundaries between the public and the private. While the idea of marriage as a relationship implies an increasing emphasis upon the private aspects of marriage, in that it is less open to some forms of public control or surveillance, the idea of the marital problem in some way implies the entry of private difficulties into the public arena. For marital problems to be talked about in these terms, the partners must have some access to some kind of public language which describes and interprets these difficulties. It is around these fluid, and often contested, boundaries between the public and the private that we find themes of risk and danger. Individual marriages may often be spoken of as 'marriages at risk', while collectively rising divorce rates may be constructed as a source of societal danger. As the public danger associated with rising divorce rates becomes apparently greater, so too does the warrant for intervention into private marriages at risk. In earlier decades, when divorce rates were lower, the individual divorced man or woman might have been seen as a source of potential danger to existing married couples (Goode 1956). Paradoxically, as the stigma of divorce decreases with growing numbers of marriages at risk, the legitimations for some form of professional intervention become greater.

In the third place, marital problems would seem, in principle, open to some kind of solution even if such a solution may actually mean the dissolution of the existing marital relationship. This implies, while perhaps not actually entailing, the presence of expertise to facilitate not only the search for such solutions but the recognition of the original condition as being problematic in the first place.

Fourthly, marital problems would seem to exist at a level deeper than their more obvious manifestations. In professional parlance, there are distinctions between 'presenting problems' and underlying conditions. This is an aspect of the 'deeper' understanding of marriage as a relationship. It is also part of the previous point, dealing with the search for possible solutions. Some kind of geological metaphor may be implied whereby certain conditions lie beneath or at a much deeper level than their surface manifestations. This deeper level may be seen in terms of patterns of communication, elaborated or developed during the history of a particular relationship, or it may point to continuities between the present marital history and previous relational histories on the part of the individual spouses, either with former partners or with members of families of origin.

Within this framework, whereby distinctions are made between presenting problems and underlying conditions, particular significance is accorded to sexuality. As we note below, the construction of 'sexual problems' has a history parallel to and inter-related with the history of 'marital problems'. Indeed, the two have been closely related and it is only relatively recently that it has been possible to talk openly and legitimately about sexual problems apart from marital problems. Yet even where the two discourses may have become more clearly separate, sexual problems may be seen as having a particular significance in the context of marital problems. Sexual problems become a major presenting problem, a major indicator of difficulties in a marital relationship.

We would argue, therefore, that while there is still value in Mills' distinction between private troubles and public issues, this distinction should not lead to the exclusion of the former from sociological analysis. Furthermore, and more importantly, we see the importance of maintaining some sense of interplay between the two. While numbers (divorce rates and their increases) have importance, public concerns about divorce are not a simple function of numerical increases. Rising divorce rates may be constructed as an important symptom of an underlying social malaise or as contributing to a whole range of other social and economic problems. Similarly, individuals experiencing the processes of separation and divorce will not be unaware of these wider debates. One way in which this particular interplay between public and private in the sphere of marriage has been understood is in terms of the 'medicalization of marriage'.

Medicalization

Within the wider context of the social construction of marital problems, it is important to take account of the particular role played by medical influences. There is evidence to show that, in various ways over the last half century, problems within marriage have been subjected to medical definition and intervention. This has by no means led to a complete medical hegemony of the domain known as 'marital problems', however, and the history of the Marriage Guidance movement in England is a useful illustration of the periodic struggles between medical definitions and those deriving from psycho-social theories. These struggles have in turn fostered differing approaches to practice, rooted in a growing division between psychotherapeutic and counselling labour. Medical thinking therefore seems to have an influence at two distinct levels: the *definition* of marital problems and the prescription of *models* of intervention. Both are of course inter-connected and each articulates in turn with the wider set of social processes that construct marital problems as a social phenomena.

Definitions From its earliest beginnings the development of psychoanalytic theory and practice created the climate in which problems within marriage could be given medical explanations. As Weeks points out, the new science of sexology had an important role in this:

> in the great classificatory zeal that produced the complex definitions and aetiologies and the new sexual types of the late nineteenth century ... we can discern the supplanting of the old, undifferentiated, moral categories of sin, debauchery and excess, by the new medical and psychological categories of degeneracy, mental illness and disease (Weeks 1981:144).

These debates about sexuality were in turn linked to population anxieties and to the development of eugenicist ideas of racial purity. From here it was but a short step to the place of sex within marriage and the issue of 'marriage hygiene'. Sexuality became, potentially at least, a source of societal danger. In the 1930s writers such as E.F. Griffith (one of the founders of the Marriage Guidance movement) were concerned with encouraging education about sex and birth control, not so much with a view to advancing the cause of sexual liberation, as with the intention of promoting the well-being of the marriage institution. Griffith believed that sexual problems were the most common stumbling block to a successful marriage, requiring education and information, both during and prior to matrimony (Griffith 1941).

The first members of the original Marriage Guidance council, formed in London in 1938, were mainly medical doctors, who saw their attempts at educating about marriage as part of a national programme of preventive medicine (Lewis, Clark and Morgan 1992). In the post-war years however, as Marriage Guidance became operational in centres throughout the country, the early model of public health quickly gave way to a curative model which drew its inspiration from psychiatry and psychotherapy. This process hinged on the growing willingness of doctors to promote medicalized accounts of the aetiology of marital problems, along with the accompanying taxonomies and classifications of marital pathology. Henry Dicks, drawing on his experience of providing marital therapy in the Tavistock Clinic, points out that in the late 1940s there was a considerable medical debate about 'marriage neurosis', a condition thought amenable to treatment using ECT (Dicks 1967:1). At that time Dicks and others were concerned with the immediate issue of rising divorce rates, which in the period after World War Two was seen as a social 'epidemic', requiring 'drastic measures of research and prevention' and as a profound threat to the family itself, portrayed in these accounts as 'the irreducible unit of social organization, rooted in biological reality' (1967:45).

It is in the context of this thinking that the 'medicalization of marriage problems becomes most visible: the application of psychoanalytical ideas of unconscious motivation to the marriage relationship: the concept of the dyad as the 'patient', the marriage as the 'sick person'; the apparently continuous relationship between the treatment of organic pathology, mental disorder and marital unhappiness (Dicks 1967:8). The explication of individual cases has an important part in this and is typically seen as an aid to both clinical practice and the training of other therapists. Marital therapy literature is thus heavily infused with the exploration of case material which depicts individual marriages receiving 'treatment' in the wider framework of concepts of and theories about marital 'psychopathology'. The following example (in which ironically the 'patients' are themselves a doctor and a nurse) is quoted in an important text on family and marital psychotherapy and provides a clear illustration:

> Dr and Mrs A., respectively trained as a physician and a nurse, were referred by another consultant for an opinion regarding the possibility of treating the marriage as well as the separate pathologies. Dr A. suffered from depressions, characterized by paranoid fantasies and homosexual fears, for which he had been treated with anti-depressant medication. His wife had suffered throughout from frigidity, associated with general emotional inhibition and fears of losing control. Previous treatment of the wife by behaviour modification techniques had been ineffective, and the sexual relationship had steadily deteriorated as the emotional problems in the marriage had grown (Teruel, quoted in Skynner 1976:125).

Such accounts, characterized by medical reductionism and the unwavering application of diagnostic labels, have several of the hallmarks of the 'clinical gaze', which we discuss in the next section. The growing interest of doctors in marital work since World War Two, albeit a very small movement within medicine itself (Clark and Haldane 1990), may therefore have had a disproportionate influence in constructing marital 'problems' as legitimate territory for clinical intervention. It has undoubtedly affected the practice of others, such as some social workers, clinical psychologists and counsellors, who have been influenced by ideas contained in the medicalized marital therapy literature or have received training in medical settings.

Models Whilst the application of medical thinking to the understanding of marital problems has been most extensively developed within the psychanalytic tradition, it is important to note a significant challenge to this model. In the early 1950s, Carl Rogers, working in the USA, had begun to develop his ideas about 'non-directive' and 'person-centred' intervention in troubled relationships (Rogers 1951). This approach rejected diagnostic techniques in favour of 'unconditional positive regard' for the client, who was to be supported through a process of self-discovery and problem resolution. 'Active listening' was the principal method in this approach in which all attempts to guide or unduly influence the client were eschewed. By the end of the decade these ideas had taken a firm hold in the world of Marriage Guidance in Britain, where they were enthusiastically championed by a new generation of marriage 'counsellors'. Counselling approaches to marital problems appeared to take hold in Marriage Guidance because they offered a non-medicalized method whereby 'lay' volunteers (after suitable training) could themselves give help to clients. This marked an important change from the volunteers earlier role as 'first aid workers' (Mace 1966:7) whose task was to 'diagnose' clients' problems before passing them on to consultant 'experts' such as doctors, lawyers, clergy. This shift from 'guidance' to 'counselling' (Lewis, Clark and Morgan 1992) set in motion important organizational changes, whereby the work of Marriage Guidance, *qua* social movement, became increasingly identified with the promotion of ideologies of counselling and personal growth, rather than the protection of the marriage institution (ibid).

This process can be seen as one aspect of the societal response to 'marital problems' which we referred to earlier. As public anxiety about divorce resurfaced in the 1960s and 1970s, attention focused less on its cause or *aetiology*, though this remained a feature of the medicalized approach to marital 'pathology' (Dominian 1980), and more on the forms of help which might be made available to those encountering marital unhappiness (Working Party on Marriage Guidance 1979). Technical divisions of labour began to emerge in the services on offer: marriage counselling, marital therapy, divorce counselling, conciliation, mediation. These in turn drew on a range of theoretical approaches: humanistic psychology, psychodynamics, systems, cognitive, rational-emotive. There thus occurred a growing differentiation within models of intervention in marriage. Though this had been inspired in part by a rejection of medical approaches, ironically it also rekindled an interest in medical models. This manifested itself in various ways. Some counsellors showed an interest in extending their practice into 'psychotherapy', i.e. working with clients over longer periods of time, exploring aspects of the unconscious and working with transference in the therapeutic relationship (Holt 1971; Mann 1974). Others developed their work as part of the primary care teams, providing counselling in general practitioners' surgeries and in health centres (de Groot 1985). In particular there was the development of a new speciality, that of sex therapy, which would take marriage counselling once again into the domain of the curative medical model.

Two clinical psychologists, writing in a recent collection on sex therapy in Britain, describe the emergence of this new field of work:

> It is perhaps surprising that something so central to marriage as sexual activity should warrant its own 'therapy' distinct from marital therapy. In the early years of psychotherapeutic endeavour marital and sexual dysfunction were both treated within the sphere of the then traditional psychodynamic approach. However, the theories of Freud and his contemporaries were developed at a time when knowledge of the physiology of sexual response was exceptionally scant; therefore these theories have been found wanting in several respects. The important breakthrough in the treatment of sexual dysfunction marked by Masters and Johnson (1970) was heralded by a growing understanding of the physiology of the sexual response. This sex therapy in opposition to psychodynamic formulations was heavily based on behavioural treatment principles (Street and Smith 1988:204).

Under-pinned by a theory and social epidemiology of sexual function and dysfunction and accompanied by a plethora of behaviour modification techniques and outcome measures, sex therapy thus made its own special knowledge claims. Its adoption within the Marriage Guidance movement constituted a major departure from humanistic, Rogerian approaches and re-introduced the possibility of clinical treatment models, along with, for the first time, more formal approaches to evaluation. Whereas the 'efficacy' of counselling was traditionally shrouded by conventions of confidentiality, sex therapy from the outset (Brown 1979) was developed as a research-based form of practice. In these aspects, and likewise in its use of an 'apprenticeship' approach to training, it therefore took on several of the dimensions of the medical model. Sex therapy has not in any sense 'taken over' marriage counselling, which continues as an independent activity; but it has led to some reappraisal of methods of training and of practice, particularly within the Marriage Guidance context. The rise of sex therapy nevertheless illustrates one strand of the continuing influence of medical ideas in the field of 'marital problems'. It bears testimony to the power of medical models of thinking and practice to extend into areas of social life where troubled relationships, rather than diseased bodies (or even minds), are the central issue. In this sense it is part of the wider process of 'medicalization' which has attracted the critical commentary of sociologists.

We should not ignore the possibility however that the activities of counsellors/therapists and sociologists in this field may hold several elements in common (Clark and Haldane 1990). For example, all may be concerned with the interpretation and meanings associated with troubled relationships and all use 'talk' as a vehicle for exploring this. Each makes use of specialized techniques for harnessing this talk: the 'therapeutic hour', the 'counselling session', the 'in-depth interview'. Each works to some canon of ethics and confidentiality within which it reconstitutes aspects of 'private' individual experience for more 'public' consumption. In doing so, whether it be through the promotion of particular caring services or through programmes of research, therapists, counsellors and sociologists in turn play their part in the social construction of marital problems. They therefore each contribute to forms of surveillance in which the medicalization of relationships may indeed constitute only one sub-set of some wider system of domination, control and regulation.

A counselling gaze?

One possible elaboration of the 'medicalization' thesis might, following Foucault, be in terms of a version of the 'clinical gaze' (Foucault 1973; Armstrong 1983). The metaphor of the 'gaze' has already been developed to explore aspects of gender and cultural relations, tourism (Urry 1990) and welfare policy (Lee and Raban 1988). On the face of it, therefore, there would seem to be some considerable potential in thinking about the 'counselling gaze', especially as the term might be applied to marital counselling.

We can begin to appreciate some of the ideas behind the development of the term 'the clinical gaze' if we focus on Foucault's contrast between the two questions 'what is the matter with you?' and 'where does it hurt?' (Foucault 1973:xviii). In the latter case, the pain takes on the status of a symptom, and what is visible or immediately felt by the patient becomes a sign of some underlying condition which can be discerned by the application of specialized medical knowledge. The particular (this patient and this body) is linked to the general, to other cases which have been noted and recorded. The physician is, through this accumulation of knowledge, able to say what is seen, and by so saying is able to show others what is to be seen (ibid:196).

The notion of the 'gaze' clearly implies detachment or, in Parsonian terminology, 'affective neutrality'. The gaze is dispassionate and unflinching. The qualifying term 'clinical' underlines that this is not anybody's gaze; it is a gaze which is ordered and codified through the development of professional bodies and medical power. Such power itself needs wider sanctioning and support and hence is related to wider processes of social and political development. The historical development of the clinical gaze is linked to wider concerns of the disciplining and surveillance of populations. For one thing, the control and monitoring of epidemics requires the increasing medical surveillance of individuals.

It should be noted that this development of the clinical gaze is not simply the replacement of an earlier, more holistic approach by an approach which is more partial, more concerned with cases, symptoms and specialized parts of the body or is less concerned with the whole person. The clinical gaze may be concerned with a whole person but it is a person which is reconstructed or re-assembled in a different way.

There are some further points that may be made about this more generalized idea of the 'clinical gaze'. In the first place, it would seem, there is always an element of danger or threat in the gaze. In the English language, at least, the term does not have entirely positive connotations. To 'look fixedly at' is close to starting, often a stigmatized activity. Under certain circumstances, a gaze or a stare may be seen as a violation of personal space or an assault on the person herself. This element of danger emerges at several points in Foucault's account. Thus he asks whether the gaze is not some form of violence 'upon a sick body that demands to be comforted not displayed' (ibid:84) and whether 'pain can be a spectacle?' Elsewhere he notes the possible tensions between the clinical gaze and established notions of morality and modesty; thus the use of the stethoscope allows for some measure of distance between a male physician and a woman's chest.

Where such dangers and tensions exist, there is a need to look also at the systems of control and legitimation. These legitimations are part of the development of medical power within society. But they also become elaborated in terms of the alleged benefits to the patient or to society as a whole.

Furthermore, the clinical gaze is collectively organized and controlled. In the operating theatre, for example, the potential clinician is taught how to gaze. Unlike the voyeur, this clinical gaze is usually obvious requiring no concealment and ultimately becoming its own legitimation. In contrast to the gaze of the voyeur the clinical gaze is often, if not always, implying some measure of willing participation on the part of the gazed-upon. The patient may seek medical help although the implications of this search may not always be fully appreciated or explained. In some cases, as we shall see, this process whereby there is some modification of the apparent lack of reciprocity may also entail some measure of resistance on the part of patients. At the very least, the physician–patient relationship is a *relationship* and not simply the unilateral imposition of the power of the former on to the latter.

It should also be noted that the idea of the clinical gaze is a clear example of the ironic mode in sociologizing. While Foucault makes it plain that no criticism of any particular body of medical knowledge or practice is implied, it would nevertheless seem likely that this way of describing shifts in medical perceptions and practices will not accord with the understandings of practitioners themselves. At the very least, Foucault is rejecting any straightforward linear model of the cumulative development of medical knowledge or the unambiguous triumph over ignorance and prejudice.

Turning from this more generalized discussion of the 'clinical gaze' to a consideration of counselling, especially of marital counselling, certain parallels and affinities are clear. Indeed, the 'counselling gaze' might, in some senses, almost seem to be a pure type of the 'clinical gaze'. Certainly some features of counselling may be readily recognized in the pages of Foucault:

> The observing gaze refrains from intervening: it is silent and gestureless (Foucault 1973:107).

Not that we should be surprised by these parallels. Medical and clinical encounters involve the particular deployment of conversational practices on the part of the doctor or physician at least as much as they involve physical examinations. Further, as we have seen, the idea of 'medicalization' entails the extension of a medical/clinical perspective to areas of practice and intervention hitherto existing within different frames of understanding.

The focal point of the counselling encounter may be seen as a particular kind of conversation. Foucault, while critical of the vocabulary of 'encounter' and 'doctor/patient relationship', writes of the elaboration of the 'gaze' through processes of listening and speaking, showing how the physician listens to the patient while simultaneously observing and developing medical science through the recording and describing of what is seen and heard. More ethnomethodological accounts might focus on the particular features of doctor/patient conversations, the features that both draw upon routine conversational practices while also having their own particularities (Heath 1986). Turning to marital counselling, two observers have written:

> Attending sessions of marriage counselling resembles conversations with an officially recognized stranger (Timms and Blampied 1985:58).

The qualifications, of course, are crucial. Conversations with a stranger have features which are distinguishable from conversations between intimates and the 'officially recognised' introduces further centrally distinctive features of the counselling encounter.

The gaze, in the more literal sense of the term, is by no means absent from such encounters. The spatial arrangements of the counselling encounter, while ostensibly designed to encourage freedom of disclosure on the part of the client and to avoid the threatening directness of the face-to-face gaze, may also enable the counsellor to monitor the client in a relatively unobtrusive manner. Certainly, counsellors are encouraged to read non-verbal communication and body language in particular way and in ways which are not routinely available to the clients themselves.

As with other clinical encounters, the operation of the counselling gaze involves the complex elaboration of connections between parts and the whole. Thus any aspect of the encounter, including time of arrival, bodily posture or tone of voice, may, in principle, be available as clues to the dynamics of a particular marriage and these clues may, in many cases, 'speak louder than words'. In a different context, Bloor and McIntosh note how, in the therapeutic community, single acts are routinely given new meanings and this process of reconstitution is a feature of the counselling encounter as well (Bloor and McIntosh 1990). In the case of marital counselling, the whole that is reconstituted through the encounter is not a particular version of the individual or the 'whole person', but the marital relationship. The crystallization of the idea of the marital relationship may be counted as one of the major achievements of marital therapy and counselling, deriving as it does from the continuous interplay between theory and practice. Theory, derived in a relatively eclectic way from psychoanalysis, systems theorizing and communications theories, provides a framework for the understanding of the particularities which emerge in the course of the counselling encounter confirm the centrality of the idea of the marital relationship.

Closely associated with the distinction between the parts and the whole and their reconstruction in the counselling encounter, is the relationship between appearance and reality. This is clearly illustrated in the central idea of the 'presenting problem'. Clients are identified as 'presenting' with a particular, overtly stated 'problem'. This may be a partner's infidelity, arguments about money, excessive drinking on the part of one partner or the other or acts of violence. These presenting problems are seen as only points of departure; the work of counselling consists, in part, of going beyond or beneath the initially stated problem to the underlying relational problems within the marriage. It is clear that, in many respects, the counselling encounter is a dangerous one:

> It appears that, in counselling, the private and the public are brought together in complex ways (Timms and Blampied 1985:vii).

In individual counselling, these dangers revolve around the 'secrets' that may be disclosed by the client (sexual fantasies being the paradigmatic example) and the dependencies between client and counsellor that are created partly as a consequence of these disclosures. These dangers are multiplied in the case of marital counselling proper, where partners may disclose not only to the counsellor but, perhaps for the first time, to each other. The developing of marital counselling practice, therefore, is not simply a matter of learning to read off from the complex and contradictory set of presenting problems and verbal and non-verbal signs but also of developing practices that can handle and contain the dangers that might arise from marital disclosures. These include a stress on confidentiality, a rigorous programme of selection and training of marital counsellors and the routine sharing and evaluating of counselling experiences, in short the development of a professional identity even where the individuals concerned may not be in receipt of a fee or a salary for their services.

One feature of the complex linkages between public and private in the counselling encounter which reflects on debates around the clinical gaze is to do with the links between marital problems as private concerns and as public issues. Foucault considers the role that public political concerns about epidemics played in the elaboration of the clinical gaze and, by way of illustration, Armstrong argues that tuberculosis shifted from being a disease of individual bodies and environmental neglect to being understood as a 'disease of contact and social space' (Armstrong 1983:11). It is not too fanciful to see a similar process at work in the history of interventions into marriage. At various stages in the history of Marriage Guidance we find publically stated concerns about marital problems, specifically as they are seen to be manifested in divorce rates, as representing a wider social problem of some magnitude. Indeed, as we have seen, it was not uncommon for public statements about divorce to talk of 'epidemic proportions'. Even where Marriage Guidance was not the source of such public statements, the growth of marital counselling could be seen as a response to what elsewhere was understood as a social 'disease'. To talk of divorce as having reached 'epidemic proportions' is not, as we have seen, simply to make a statement about the sheer numerical weight of individual unhappinesses. It is also to see such rates of divorce as being likely to contribute to a wider range of other social problems, as symptoms of some underlying social malaise or, indeed, as both. The history of Marriage Guidance in Britain can be seen in terms of a sometimes uneasy occupation of the space where private problems meet public issues.

Consideration of the possibilities of talking about 'the counselling gaze' should not be seen as being a simple and mechanical application of a Foucauldian perspective but as a way of entering into the complexities and ambiguities of the counselling encounter and, further, of providing additional elaborations of the idea of 'the gaze' itself. In the first place, let us consider the theme of gender, an issue somewhat muted in Foucault's own discussion and in some subsequent developments of the idea of the clinical gaze. Apart from a covert recognition of a male physician with male or female patients (as the discussion of the stethoscope illustrates) there is little recognition that the participants in a clinical encounter have gendered identities. In the case of marital counselling this comes to the fore in a variety of ways. In the first place, in Marriage Guidance, we have the not uncommon phenomenon whereby a largely female 'technical' staff work for an organization which is managed largely by men. This is sometimes a popular theme in members' accounts of Marriage Guidance (where the contrast may sometimes be expressed in Jungian terms) and counselling skills may be constructed as the development of particularly feminine skills or the elaboration of 'caring work' or 'emotional labour' (James 1989). Secondly, the clients themselves have gender identities and issues around what might be appropriate for women and for men in a marriage might be central themes in the counselling process. Perhaps of particular relevance here is the debate around whether men have greater difficulties than women in questions of disclosure and whether, perhaps as a consequence of this, they have different expectations about the counselling encounter than their female counterparts (Brannen and Collard 1982; Hunt 1985; O'Brien 1988). It is maintained that men are more likely to seek identifiable solutions and positive advice rather than a more diffuse airing of problems. Thus the operation of the counselling gaze may have different connotations for men and for women. Finally, the counselling encounter is a gendered encounter with all the dangers that this may imply; dangers which are recognized in those particular modes of counselling where a heterosexual couple are matched by two counsellors, one woman and one man. In this way, perhaps, assumptions about gender and sexuality are reproduced through the counselling encounter.

Another feature which is muted, if not entirely absent, in the Foucauldian account is the idea of resistance. The gaze as a metaphor implies one-way surveillance. However, as we have suggested, some measure of interaction and reciprocity is entailed and this may include a measure of resistance:

> The patient is not simply an object which will conform; it possesses, in its potential for default, a germ of idiosyncracy (Armstrong 1983:103–4).

In the case of Marriage Guidance, signs of resistance on the part of the clients may be seen in the expression of negative feelings about the counselling received or about the counselling process in general, or in ending the counselling process by default, either through cancellation or through simply failing to turn up (Hunt 1985; Timms and Blampied 1985). Certainly the picture that emerges of the client in the few studies that we have carried out, is not one of individuals or couples submitting passively to a counselling gaze. It may be suggested, indeed, that there may be an element of rational calculation on the part of the clients where resort to counselling is understood to be a legitimating device to demonstrate to the world at large that every alternative had been explored prior to a decision to separate. To a large extent, of course, these possibilities for client resistance reflect the ways by which clients are recruited into the counselling process (Brannen and Collard 1982).

To reflect, albeit briefly, on the different understandings that clients may have of the counselling situation is also to raise the possibility that the perceptions of the counsellors might also be more heterogeneous than any sociological model might suggest. There is always a problem in matching constructions based upon surveys of the literature, in this case counselling literature, and constructions based upon members' accounts. Timms and Blampied, for example, in their small-scale study of counsellors, found that, contrary to the arguments of Halmos (Halmos 1965), counsellors did not see themselves as applying science, did not suffer any great confusion between influence and directiveness and did not have firm optimistic beliefs about the triumph of love over hate (Timms and Blampied 1985:31). As we have seen, the history of Marriage Guidance does show a complex range of differences and contradictions on the part of counsellors and between counsellors and managers, and calls into question any straightforward picture of the steady growth of the counselling gaze (Lewis, Clark and Morgan 1992). One example of these differences concerns the 'title' of those who attend counselling sessions. The term 'patient' was offered, but rejected, in the early days and there continue to be some doubts about the use of the term 'client'.

All this suggests some limitations in adopting, uncritically, the idea of the 'clinical gaze' to the study of marital counselling. It may be noted that Foucault was often writing about the teaching hospital where the clinical gaze is, as it were, amplified in the context of the operating theatre. Superficially, it may seem difficult to see what the equivalent of the operating theatre might be in the case of marital counselling, especially given the norms of confidentiality. But the gaze can be seen in a more developed form in family therapy's use of the one-way screen, video recording, live supervision etc. And, indeed, detailed accounts of anonymized cases may provide some kind of functional equivalent, even in the case of marital counselling (Freeman, 1990).

The counselling model, far from being a simple extension of or development from a clinical model might, in some circumstances, come into conflict with the clinical model. For example there may, in principle at least, be conflicts as to whether a certain 'disorder' in marriage (such as schizophrenia or alcoholism) is organic in origin or a product of the relationship itself. As we argued in the previous section, there are possible tensions and overlaps within marital counselling itself, as it draws eclectically upon a range of traditions including Rogerian non-directive models and the more behaviourist traditions which inform much sex therapy. For example, the following story appeared in *The Independent*:

> Loss of desire is now recognised as the commonest sexual problem for both men and women, according to Relate – formerly the Marriage Guidance Council. In most cases the cause lies in a couple's relationship or in past traumas, but increasingly it is found to be the result of hormone problems (*The Independent* 9.11.90).

While an understanding in terms of relationships or past traumas and an understanding in terms of hormone deficiencies may both be examples of a clinical gaze, these are really quite different kinds of gaze and these differences should be recognized and elaborated.

One final objection needs to be considered. People receiving marriage counselling, marital therapy or marital sexual therapy are not generally forced to attend such sessions. The main legitimation for such manifestations of the clinical gaze is that the 'clients' willingly subject themselves to this gaze and hence the links with notions of surveillance at a societal level are, at the very least, a little premature. This is true; but it should also be recognized that here, as elsewhere, there is some margin of doubt as to the extent to which an action is 'freely' chosen. Couples may attend marital counselling as a consequence of pressures, real or imagined, from one of the partners or from kin, friends or members of the 'caring' professions. Certainly, wherever there is felt to be some measure of stigma attached to divorce or separation or where the marital relationship is valued as the single most important adult relationship, the degree to which the resort to marital counselling is freely chosen must be open to question.

The marital gaze

It is tempting, indeed possible, to see the growing 'concerns' about marriage and divorce to be little more than a reflection of wider tensions and contradictions surrounding the nature and scope of state intervention in a late capitalist society. In particular, emphasis might be given to the elaboration of a 'New Right' philosophy, stressing both the limitation of state expenditure and the extension of individual moral responsibilities. Thus, in order to reduce the short-term and possible long-term costs of single-parenthood on both the state and the wider society, there is a need for the encouragement of a more 'responsible' attitude towards marriage and parenthood. In this encouragement, a particularly important role might be allocated to the work of voluntary organizations such as Relate. While the support of such organizations might entail some relatively modest state expenditure, this may be seen as being money well spent; certainly some of Relate's recent publicity would seem to be arguing along these lines.

We see such wider economic and political considerations as being important strands in the overall argument but not as constituting the whole story. For one thing, we see the process whereby marital problems are constructed and understood as being bound up, in complex ways, with the growth of wider understandings or marriage in relational rather than in institutional terms and the growth of professional or quasi-professional discourses around the nature of marriage. We see some measure of convergence or overlap between wider processes of medicalization, the overall elaboration of a counselling gaze in a wide variety of organizational and inter-personal settings and the construction and understanding of marriage in broadly relational terms. This nexus of cross-currents and interacting influences may indeed be placed in the context of an analysis of state and economy in late capitalist society. However, we see tensions and contradictions, and any unity that may be said to exist must be seen as a complex unity that includes and takes account of such contradictions.

It is clear that private marriages at risk may also be seen, collectively, as a potential source of societal 'danger'. While the more neutral-sounding, perhaps more medicalized, terminology of 'risk' and 'danger' may replace more overt moral or religious disapproval and while it may be claimed that divorce has been 'destigmatized', it is clear that the notion of 'marital problems' continues to constitute a powerful discourse on the borders of the public and the private. This theme of danger is, implicitly, given greater emphasis in the controls and confidentialities which surround professional interventions into marital disorders'.

'The marital gaze', therefore, is composed of a variety of social and historical processes. There is the growth of a medical discourse around marriage and sexual relationships – a discourse which has not, however, been unchallenged. There has been the development and modification of a generalized 'counselling gaze' in the direction of a focus on marriages or marital problems. And there has been the growing and increasingly pervasive understanding of marriage in relational terms, with the consequent particular and special understanding of marital problems. In this process, in this constant interplay between private problems and public issues, it is likely that even those couples who have never been anywhere near a marital counsellor may have developed some measure of self-surveillance. Thus individuals and couples learn to monitor and gaze upon their own relationships in a particular way. This may be the major triumph of such processes: the development of a 'marital gaze'.

This chapter is based upon research funded by the ESRC under the title 'Marital Problems and the Marital Agencies Since the 1930s' (Ref. No. R900 231060), conducted by Jane Lewis, David Clark and David Morgan. A book *Whom God Hath Joined Together: The Work of Marriage Guidance, 1920–90* based on the project was published by Routledge in 1991. We should like to thank the ESRC for funding the project, Relate for providing access and assistance at all levels and the editors of this present volume for their help and encouragement.

References

Armstrong, D. (1983), *The Political Anatomy of the Body*, Cambridge: Cambridge University Press.

Bloor, M. and McIntosh, J. (1990), 'Surveillance and concealment: a comparison of techniques of client resistance in therapeutic communities and health visiting', in Cunningham-Burley, S. and McKeganey, N.P. (eds), *Readings in Medical Sociology*, London: Tavistock/Routledge.

Brannen, J. and Collard, J. (1982), *Marriage in Trouble*, London: Tavistock.

Brown, P.T. (1979), 'Practical modifications of Masters and Johnson's approach to the treatment of sexual dysfunction', unpublished Ph.D. thesis, University of Leicester.

Burgoyne, J. and Clark, D. (1984), *Making A Go Of It*, London: Routledge and Kegan Paul.

Clark, D and Haldane, D. (1990), *Wedlocked? Research and Intervention in Marriage*, Cambridge: Polity Press.

Dicks, H.V. (1967), *Marital Tensions*, London: Routledge and Kegan Paul.

Dominian, J. (1980), *Marital Pathology*, London: Darton, Longman and Todd/BMA.

Durkheim, E. (1951), *Suicide: A Study in Sociology*, Glencoe, Illinois: The Free Press.

Foucault, M. (1973), *The Birth of the Clinic*, London: Tavistock.

Freeman, D.R. (1990), *Couples in Conflict*, Milton Keynes: Open University Press.

Goode, W.J. (1956), *Women in Divorce*, New York: The Free Press.

Griffith, E.F. (1941), *Sex and Citizenship*, London: Gollancz.

Groot, M. de (1985), *Marriage Guidance Counsellors in the Medical Setting*, Research Report No. 1, Rugby: National Marriage Guidance Council.

Hall, S. et al (1978), *Policing the Crisis*, London: Hutchinson.

Halmos, P. (1965), *The Faith of the Counsellors*, London: Constable Heath Co.

Heath, C.C. (1986), *Body Movement and Speech in Medical Interaction*, Cambridge: Cambridge University Press.

Holt, N. (1971), *Counselling in Marriage Problems*, London: National Marriage Guidance Council.

Hunt, P.A. (1985), *Clients' Responses to Marriage Counselling*, Research Report No. 3, Rugby: National Marriage Guidance Council.

James, N. (1989), 'Emotional labour: Skill and work in the social regulation of feelings', *Sociological Review*, **37** (1): 15–42.
Lee, P. and Raban, C. (1988), *Welfare Theory and Social Policy*, London: Sage.
Lewis, J., Clark D. and Morgan, (1992), *Whom God Hath Joined Together: The Work of Marriage Guidance, 1920–90*, London: Routledge.
Mace, D. (1966), 'Marriage counselling in the USA', *Marriage Guidance*, January, pp 7–10.
Mann, A. (1974), *The Human Paradox*, Rugby: National Marriage Guidance Council.
Masters, W.H. and Johnson, V.E. (1970), *Human Sexual Inadequacy*, London: Churchill.
Mills, C.C. (1959), *The Sociological Imagination*, New York: Oxford University Press.
Morgan, D. (1990), 'Institution and relationship in marriage', unpublished paper, Madrid: World Congress of Sociology.
O'Brien, M. (1988), 'Men and fathers in therapy', *Journal of Family Therapy*, **10**: 109–23.
Rogers, C. (1951), *Client-Centred Therapy*, Boston: Houghton Mifflin.
Skynner, R. (1976), *One Flesh, Separate Persons*, London: Constable.
Street, E. and Smith J. (1988), 'From sexual problems to marital issues', in Cole, M. and Dryden, W. (eds), *Sex Therapy in Britain*, Milton Keynes: Open University Press.
Timms, N. and Blampied, A. (1985), *Intervention in Marriage: The Experience of Counsellors and Their Clients*, University of Sheffield, Social Services Monographs: Research in Practice.
Urry, J. (1990), *The Tourist Gaze*, London: Sage.
Weeks, J. (1981), *Sex, Politics and Society*, London: Longman.
Working Party on Marriage Guidance (1979), *Marriage Matters*, London: HMSO.

7 'To hell with tomorrow': Coronary heart disease risk and the ethnography of fatalism

Charlie Davison, Stephen Frankel and George Davey Smith

Introduction

The new General Practitioner contract, jogging, High Street cholesterol screening and the stacks of 'low-fat', 'high-fibre' products on supermarket shelves are all surface indications of a major shift in the foundations of the way British society deals with health and illness. Over the last two decades, the traditionally recognized advantages of prevention over cure have finally made the transition from wise old saw to official health policy; and nowhere more completely than in the field of circulatory disorder.

In terms of the individual's interaction with medicine, the development of behavioural prevention entails the extension of a cultural system largely involved with diagnosis and treatment into the greyer areas of prediction and avoidance. This is an issue which raises serious questions concerning not only the role of the caring professions, but also the conceptual position of health itself. If the integration of disease prevention into everyday life is to be successfully achieved, the magnitude and import of the cultural changes afoot need to be recognized. The findings of the Cultural Aspects of Ischaemic Heart Disease research project[1], which are partly reported here, throw some light on the complex and sophisticated ways in which medical and quasi-medical advice on behaviour change is evaluated by the public at large. Here we report on the popular understanding of the causes and distribution of heart disease and show how the relevance of adopting 'healthy lifestyles'[2] is socially patterned.

We make use, here, of two quite different types of data.[1] On the one hand, the analysis draws on the months of observation and informal interaction between the researcher (CD) and the residents of the three fieldwork areas. Contacts were made primarily via the random sample (selected individuals, their kin, household members and friends) but also more opportunistically (individuals encountered in shops and pubs, friends of friends). On the other hand, it draws on the verbatim transcripts of the taped semi-structured interviews with randomly sampled informants. Combining formal and informal, quantitative (in a small way!) and qualitative, survey and ethnographic data has been a specific goal of the research project.

Lifestyle and coronary heart disease

Coronary Heart Disease (CHD) is the medically registered cause of about 25 per cent of all deaths in the United Kingdom. CHD morbidity within the British population is also extremely widespread. This epidemiological situation means that CHD, as a type of 'heart trouble',[3] is frequently encountered by members of society in their personal social networks. Knowledge and attitudes relating to the causes of CHD and its possible avoidance are, thus, not only based on information communicated through professional, official and media channels, but also stem from the collective observation and socially-based analysis of real-life phenomena. It is this cultural process of explanation and appreciation that is our main concern here.

As society's most widespread multi-factorial chronic disease, CHD has attracted intense interest among the individuals and institutions involved in the preventive medicine movement. Epidemiological and clinical research over several decades has identified dozens of behaviours and conditions associated with the onset of the disease, many relating to personal lifestyle, and many relating to other fields of life. While strong debates exist within the academic and professional disciplines of epidemiology and public health medicine concerning the relative weights of different risk associations, health educators and product marketing professionals have waged an intense public campaign to place lifestyle at the centre of CHD causation.[4]

Blaxter (1979) and Cornwell (1984) have, amongst others, illustrated that British people commonly operate an explanatory framework which regularly includes a moral or quasi-moral judgement of the degree of self-infliction involved in an illness episode. Different disorders, in this sense, tend to have different public images in terms of the likelihood that sufferers have 'brought it on themselves'. While some degree of self-infliction may be discerned in almost any illness (or other misfortune), certain ailments are much more strongly and regularly linked to lifestyle causes than others. Diabetes, for example, along with some cancers, brain haemorrhage, and many common infectious diseases generally stand at the 'least fault' end of the spectrum. Sexually transmitted diseases, cirrhosis of the liver and lung cancer are more strongly associated with a personal contribution towards cause from the sufferer.

In this respect, CHD is in an interesting position. As we have discussed elsewhere, a central element in lay explanations of cases of heart trouble is the idea of 'candidacy' (Davison et al 1991). On the one hand, this notion involves the putting together of various personal attributes in the construction of an idealized profile of the CHD sufferer, a profile which includes important elements based both on lifestyle and non-lifestyle issues. On the other hand, the candidacy system for explaining cases of CHD also involves the recognition that the illness can visit a sufferer both suddenly and for no apparent reason. This close interaction between the explicable and the idiopathic in the popular explanatory culture of CHD makes it a particularly appropriate field for an investigation of so-called fatalistic perspectives.

In the field of heart disease, the current orthodoxy of prevention is centred on the need to change daily habits, particularly those involving food, tobacco and exercise. The strong implication flowing from the content and style of much healthy lifestyle advice is that an individual can avoid heart disease through the triumph of self-control over self-indulgence. Our data, and the much larger national Health and Lifestyle Survey (Blaxter 1990) indicate that a knowledge of the three main medically-identified behavioural risk factors for heart disease is extremely widespread throughout all age, gender and socio-economic status groups. It is also clear, however, that knowledge of a behavioural cause of ill health does not automatically imply the abandoning of that behaviour (Blaxter 1990; Dean 1984). Given that good health is an almost universally desired goal, one somewhat crude response from professional health education has been to question the rationality of some sectors of the lay population and label them 'fatalistic'.[5,6]

A more sophisticated analysis has identified the existence of cultural norms which present obstacles to behavioural change. These include culinary traditions of high saturated fat consumption, the strength of association between exercise and youth, the importance of smoking and drinking in affirming the identity and belonging of the social self, and the cultural complex which links behavioural risk with ritual occasion and group enjoyment. It was partly in response to this analysis that traditional health education (the simple transfer of information) gave way in the 1980s to attempts to engineer new social norms and a new environment of choice and thus became modern 'health promotion' (Tones 1987).

From the point of view of medical social anthropology, however, this interest in culture appears somewhat imbalanced. It concentrates almost exclusively on cultural norms relating to social behaviour and pays scant regard to the crucial area of explanatory culture. By this we mean the social and intellectual processes which account for and make sense of illness causation and distribution as they appear in everyday life. It is to this second area of popular medical culture that we now turn. Our discussion consists of three parts: first the placing of lifestyle in a complete conceptual landscape of pathogenesis and salutogenesis; second an examination of beliefs and explanations concerning the random distribution misfortune;

third the social distribution of these ideas. Our overall aim is to illustrate the importance of notions of randomness, luck, fate and destiny in constructing and maintaining rational behavioural choices in the context of the 'lifestyle revolution'.

Healthy lifestyles in context

Voluntary behaviours (lifestyles) are by no means the only determinants of chronic disease recognized by popular British medical culture. Other facets of life which, in the perception of our informants, have a long-term effect on individual health status but cannot be individually controlled in any obvious way, can be grouped, for the purposes of analysis, into three interlocking fields:

(i) Fields involving the self-evident personal differences between individuals (e.g. heredity, upbringing, inherent traits).
(ii) Fields involving the social environment (e.g. relative wealth and access to resources, hazards associated with occupation, loneliness, involvement in accidents).
(iii) Fields involving the physical environment (e.g. climate, natural dangers and disasters, environmental contamination).

While we follow, here, the roughly orthodox path of placing an analytical or categorical division between lifestyle on the one hand and three 'non-control' fields on the other, the conceptual landscape we have encountered in our interactions with the general public contains no such hard and fast barrier. Rather, each element of the three fields listed above has a more complex relationship with individual health status and individual participation in healthy lifestyles than the control/non-control dichotomy implies. In each case there is, on the one hand, a direct influence on health status, and on the other, an indirect influence, mediated through an effect on the possibility, ease and worth of voluntary participation in healthy lifestyles. The model can best be illustrated with an example. In the popular explanatory culture of IHD, being a 'worrier by nature'[7] can have various direct effects on the health of the individual. These are caused by the clenching of internal organs and muscles ('always being tensed up'), the disruption of normal blood chemistry ('too much adrenalin') and the consequent disruption of regular heart beat. Over and above these mechanisms, there exists the possibility of indirect health damage brought about through an effect of the condition on lifestyle. In the case of the 'worrier by nature', this occurs through combatting worry with tobacco and alcohol and allowing worry to dictate inappropriate eating habits.

Being a 'worrier by nature' belongs to the first field in the simple schema listed above: the realm of heredity and inter-personal difference. Worry, anxiety, stress etc. can stem from either of the other two fields, of course. The relationship of the condition which produces worry to lifestyle, however, is the same in each case. Worry impacts on lifestyle in the same way irrespective of whether it springs from personal nature or from debt, bereavement, family crisis, unemployment etc.

Because the discourse of preventative medicine and health promotion makes such play of personal behaviour as the central field of public health, it is worth making the observations of our research on the conceptual position of lifestyle in popular health culture absolutely clear: As determinants of health status, the formal or structural position of individually uncontrollable conditions and the ostensibly controllable lifestyle fields is the same in every case. Health status is related to the condition in two ways: a) In a direct fashion, 'in its own right'; b) Indirectly via the condition's influence on lifestyle, which in turn influences health status. Thus any voluntary change in behaviour is always weighed up in the context of the given situation. It is because of the operation of this simple model that giving up smoking is a different proposition to someone who works in the polluted atmosphere of a petro-chemicals plant and someone who spends their day in an air-conditioned suburban office block.

The unique combination of factors influencing each individual life (conditions running in the family, circumstances of upbringing, disposable income, personal nature, past exposure to environmental hazards, significant life-events such as bereavement or divorce etc) dictates that decisions concerning lifestyle change are both complex and highly personal. Lifestyle, indeed, does not really exist as an independent category. Rather it is only articulated in terms of its intimate but varied relationship with the other elements in the model.

Because diet, exercise, and drug consumption cannot be seen as being independent of the social and economic structures within which behaviour (and culture) are sited, they must also be involved with the realities of relative inequality in a stratified society. As an issue concerning voluntarism and the possibility of individual choice and freedom, the relationship between health and lifestyle is of evident political importance. This political issue has resided at the centre of contemporary professional debates concerning the social distribution of mortality and morbidity since the publication of the Black Report (Townsend and Davidson 1982).

The intimate relationship between health, behaviour and circumstance leads us to propose that there exist important inequalities in the relevance of lifestyle change for individuals and groups in society. By this we mean that the construction of beliefs concerning the desirability, possibility and efficacy of lifestyle change takes place only against the back-drop of an assessment of the individual's position in relation to other salutogenic and pathogenic fields. Inequalities in the assessment of relevance exist both where belief in the benefit of lifestyle change is concerned and in actually adopting healthy lifestyles. It is our observation that both believing in the benefits and adopting the changes are easier and more common amongst those who feel it might work, have observed it working in others like them or have tried it and found it to work. The findings of the Health and Lifestyles Survey (Blaxter 1991 esp. Chap. 7) also tend towards this conclusion, suggesting a clear relationship between relative social and environmental advantage and benefit from following healthy lifestyle advice.

General principles and individual events – An explanatory role for 'fatalism'

The importance of understanding the full conceptual landscape of control and lack of control in the health field lies in the fact that prevention implies knowledge about cause. Attempts to exercise control over the timing and nature of events presupposes the existence of an explanation or set of explanations which account for the occurrence and distribution of the events themselves. It has long been a commonplace observation in the discipline of social anthropology that cultural systems of explanation or accountability need to address two distinct issues. In the first place the general kind of misfortune requires explanation: how and why does it happen? In the second place, the site and time of particular misfortune require explanation: how and why did it happen to this person at this time? (Evans Pritchard 1937).

The healthy lifestyles movement within preventive medicine represents an attempt to reduce the second area of explanatory culture to a subset of the first. Indeed, as Gifford (1986) points out, the central concept of prevention (that of 'risk') is actually produced by the transposition of epidemiological association derived from whole populations to the field of an individual life. But the particular refinement of the healthy lifestyles movement as applied to heart disease has been to attempt to place one set of general epidemiological principles (those involving behaviour open to personal control) at the centre of explanatory culture.

As we have pointed out elsewhere (Davison et al 1991), an integral part of the modern prevention movement has been the public lowering of the threshold of risks concerning personal lifestyle. Although this process is ostensibly involved in extending the bounds of explicability and control, it has the paradoxical effect of highlighting anomalies. This has happened because aspects of life hitherto considered normal and safe (chip eating, lounging about etc) have become labelled as pathogenic. Because so many other factors are involved in illness causation and distribution, a result of this ideological development is that the number of people who survive dangerous lifestyles and do not get ill grows. Simultaneously, the cases of individuals who do all the 'correct' healthy things and yet still succumb to heart trouble become very well known (Davison et al 1991). Because these anomalous deaths and unwarranted survivals are noted, the tension or conflict between general principles and individual events is constantly made plain to members of society and has to be dealt with. An extremely common explanation of an anomaly thus becomes 'it was just one of those things', the full but unspoken version of which is 'it was just one of those things which violates general principles'.

Accounting for the randomness and scatter that exist around the epidemiological trends is not a central issue for public health professionals. Rather, they deal with the trends themselves and concern themselves with taking action directed at amending a probabilistic future. Popular health culture, on the other hand, cannot turn its back on any illness or death. Those which violate general principles must also be explained. It is within this context that an ethnography of fatalism is important, as it seeks to throw light on the cultural structures within which common, but apparently anomalous, events can be accommodated. As they include many variations on the theme of 'fate', these areas of cultural life could fairly accurately be termed 'fatalism'.

Randomness, chance, luck and fate in everyday explanations of illness and death

In the everyday discourse on health and illness which forms part of normal social life, it is extremely common for people to make references to luck. Often, this takes the form of the simple observation that someone has 'been lucky' or 'unlucky' in avoiding or succumbing to an ailment of one type or another. These phrases are often used when the timing of an illness episode is under discussion (during a holiday, at the time of a child's birthday etc.). Such regular and brief mentions of luck and fortune could be nothing more than a conventional linguistic usage, and not be particularly meaningful in a detailed or substantive way. It is our assessment, however, that this particular aspect of normal language is an indication of an important area of thought and explanation concerning both the cause and distribution of heart trouble and other serious ailments.

In the case of the onset of serious life threatening illness or sudden death, simple references to luck are regularly supplemented by ideas which belong to more detailed styles of explanation. We have observed three distinct areas within the general field of luck, fate or destiny: religious ideas, notions of pre-destination and allotted time, and the missile analogy. The first area of non-material explanatory belief is concerned with religious ideas and the existence of a deity who exercises an overall control of human life. The operational idea here is that God has a general design of all life, which encompasses the beginning and the end of individual human existences. Exactly when and how a life will end, and the occurrence of events and misfortunes within it, are seen as phenomena which fit the divine plan, rather than the personal designs of individuals themselves. People who could be described as seriously religious, indeed, indicated a certain distaste for some aspects of the healthy lifestyles movement, being offended by the presumptuous and arrogant idea that individuals can exercise control over the timing and nature of their own death.

In our formal interviews and informal observations, the notion of a divine design appeared as references to 'God's plan', 'when the Almighty decides', 'it was God's will' etc. A recurrent theme was that an individual is 'called' by God when, where and how the plan dictates. Such events are, by their nature, unpredictable and uncontrollable at either an individual or collective level. They are also inexplicable in terms of other aspects of material life, concerning as they do a deity who is often recognized as 'moving in mysterious ways'. Some individuals were observed to make links between individual behaviour, divine judgement of that behaviour and the occurrence of illness and death. Health beliefs concerning divine retribution were uncommon, although the image of a vengeful God appeared to have an appeal to some individuals who combined an apocalyptic view of HIV/AIDS with deep prejudices concerning sexual behaviour.

The notion of 'being called' finds an accommodation with a second area of non-material or supernatural theorizing about the occurrence of illness and death. This concerns, again, the idea of each individual having an 'allotted time' for life. Although a biblical reference ('three-score year and ten') is one of the most common phrases used in this connection, the idea of an allotted life-time is not necessarily a religious one. Non-religious images are also employed, the most common being the 'grim reaper' whose hour glass indicates that an individual is somehow 'ripe' for death. Other common images include life being analogous to the time that a boat is hired for pleasure ('come in number 23, your time is up') or subject to the oversight of an actuarial figure 'with a book' who calls names or numbers as they become due. The implication of these ideas is that a person will die 'when their time has come', irrespective of their participation in healthy lifestyles. These essentially non-religious theories of pre-destiny in illness and death are most commonly voiced in everyday life as explanations which make an appeal to a general idea of the fate or destiny of an individual, because that is the way the world works. The idea is generally expressed in a short sentence which is made up of the repetition of two clauses: 'what's for you is for you', 'if it's gonna happen, it's gonna happen' or, in the words of an oft-quoted popular song, 'Que sera, sera. Whatever will be, will be'.

The missile analogy also uses the idea of illness or death coming to an individual by way of their name or their number. In fact it can be seen as a sub-category of allotted time, in that it concerns a particular method of time being 'called'. The general notion is that these events fly through the air, targeted at specific individuals, whose names or numbers are written on them. Thus a heart attack will 'get you' if and when there is one flying around 'with your name (or number) on it'. This particular image was connected by several informants to times of war and the ways in which combatants and bombarded civilians sought to explain the random distribution of sudden and violent death:

> *Informant*: 'Well I suppose I'm more of a fatalist because when I was in my teens the war was on and in the bombing and everything else you didn't know who it was going to hit, so you grew up with the idea that it was going to happen. If it's got your name on it it'll hit you, and that's my attitude now. If it's got your name on, it'll hit you. Whether its God, fate, call it what you like. I don't know. Destiny.'

While it is commonly believed that the missile analogy dates from the trenches of the First World War, the antiquity of the general idea in British culture is indicated by the Prince of Denmark's famous reference to 'the slings and arrows of outrageous fortune' (*Hamlet*, III. i.). It should be noted here that the idea of luck varies in each of the pathogenic/salutogenic fields described earlier. While a general conception of luck runs through much explanatory culture, it is evidently more important in some elements than in others. Luck is used to make sense of many aspects of the way that individual lives articulate with collective events. The category finds as much use in explaining being born into a family with weak chests (while others were not), living in a valley which received a downpour of fall-out laden rain over the Chernobyl weekend (while other valleys did not) or being pre-destined for death in a train crash (while others are 'booked in' for a heart attack). It is also of central importance in explaining the apparently patchy effects of lifestyle and thus accounting, for example, for heavy smokers who do not get bad chests and nonsmokers who do.

The strong presence of a belief in luck in explaining the distribution (if not the actual mechanisms) of medical misfortune is paralleled by the ubiquity of games of chance in modern British society. Sports gambling, fruit machines, draws, raffles, premium bonds and lotteries are valued activities for much of the population and the idea of 'the luck of the draw' is an important concept in making sense of many aspects of life.

Attempts to predict and/or manipulate the workings of luck are also common, and many individuals engage in them without necessarily thinking of themselves as particularly superstitious people. Amongst the activities aimed at the manipulation of luck in general (and so to some extent towards luck in the field of illness avoidance) which were encountered during fieldwork included: touching wood when mentioning misfortune, crossing fingers while awaiting the outcome of an important process, refraining from discussion or mention of misfortune in order not to 'tempt fate', wearing lucky clothes or adornments, carrying out daily personal rituals (pocket tapping, ornament touching etc.), tossing coins into ponds and fountains, observing certain animals (magpies, crows, black or white cats), and the use whenever possible of particular numbers or colours. Many informants were observed to pay close attention to 'their stars' in the astrological columns of newspapers and magazines and some had taken part in various types of fortune telling. This widespread interest in the prediction and manipulation of luck and destiny suggests that they are agents of causation and distribution which are not always seen as entirely outside the control of the individual.

The social distribution of ideas about luck and fate

Analysis of the formal, semi-structured interview material gathered from the random sample in each fieldwork area showed that luck, fate and randomness were discussed by: 42 per cent of all respondents, 43 per cent of respondents classified as having non-manual occupations, 42 per cent of respondents classified as having manual occupations, 45 per cent of male respondents, 40 per cent of female respondents, 23 per cent of respondents aged 18–30, 47 per cent of respondents aged 61–75, 35 per cent of respondents resident in Plasnewydd (Inner city Cardiff), 44 per cent of respondents resident in Porth (Rhondda town), 48 per cent of respondents resident in Llangammarch (Rural area, Brecknock). These figures indicate that there may be two important aspects of the social distribution of the perception of the importance of randomness, chance, luck, fate, destiny etc. in accounting for illness. In the first place, it is clear that older people are much more likely to mention the category in the context of a formal interview with a researcher from the official health world. This could either indicate that the older informants had ideas belonging to a previous era and that the younger were more 'modern' (and less fate-oriented) in their outlook, or that older people are more experienced and wiser and that today's young will also feel more strongly about luck in relation to illness when they have seen a bit more of life. In the second place there is an indication that ideas concerning luck, randomness etc. have a greater currency in the rural area, are in an intermediate state in the industrial town and are relatively less important in the inner-city neighbourhood.

Analysis of the contents of comments made during interviews throws little light on the age question, but shows important differences between regions. In general there was little substantial difference in opinions and beliefs held by people in different socio-economic groups. It was observed, however, that ideas in this field tended to be quite clearly embedded in the personal experience of each individual. A farmer, for example, was of the opinion that: 'People are just like animals, isn't they? Some's born stronger than others and that's that.' Similarly, miners, ex-miners and other residents of the Rhondda study area made regular reference to the caprice of accident and disaster in the pits when discussing chronic illness and sudden death. Several women drew attention to the apparently random distribution of and relatively 'difficult/easy' labour and childbirth as an indication of the 'no rhyme or reason' theme.

One particular aspect of the content of interview comments showed strong regional or area variations which echoed the simple percentage counts listed above. This was the number of people who supported their belief in luck, fate or some other random principle with observations of specific cases. These cases took the form of explicit comparisons between people known to the informant who had become ill unexpectedly (that is to say anomalously as far as general principles are concerned) and people who could be expected to suffer illness, but enjoyed good health. A forty-four year old male forestry worker commented:

> *Informant*: 'I think there is some, some amount of luck innit.'
>
> CD: 'Yeah yeah.'
>
> *Informant:* 'Like there's my mate in the village here like uh, well he was uh a perfect uh fit well, he use to be on the forestry only he was driving uh a bull dozer type thing you know, and he was quite slim and he enjoyed the drink, he, of course he smoked a bit and and what is he forty-six, he just... He was coming home from work one night, he had this terrible pain and he went to hospital, he had a heart attack and died.'
>
> CD: 'Really?'
>
> *Informant*: 'Like uh, he was a chap, he wasn't red faced, he looked a normal fit person like, and he just went like that. Well, well is that ... is that unlucky or what? Is it, like? Innit. Like if he'd had a warning you know well. I know he smoked yeah but uh like there's other people, there's other people smoke like twice as much as him and live twice his age.
>
> CD: 'Mmm, mmm. So there was no ... .'
>
> *Informant*: 'It's very odd like innit, its funny.'

Similarly, a 56 year old male local government senior officer reported:

> I have never smoked, I've never drunk to excess to that extent but I got a heart attack but then you see people doing all these things and they're completely healthy so it makes you think about your health. Why me and not them?

This type of story belongs to the practice we have termed 'lay epidemiology', where cases of illness and death from personal observation, histories known through personal and kin networks, media reports etc. are discussed and analysed. Part of this process is involved with the collection of 'evidence' or 'data' which can be used to support or challenge suspected aetiological processes (Davison et al 1991).

While this type of support for opinions concerning randomness and luck showed no significant variation by age, gender or social class, the inter-area variation was striking. In Plasnewydd (inner city) only two informants of the twenty two who mentioned randomness supported their ideas with personal observations. In Porth (Rhondda), eleven of the twenty four mentioning randomness used personal observations. In Llangammarch (rural Brecknock), the figures were fifteen out of twenty eight.

'Fatalism' and lifestyle

One of the striking aspects of the facet of popular explanatory culture which deals with randomness, luck and chance, is that it finds an easy accommodation with other, more mechanistic beliefs about the cause and possible avoidance of illness, including lifestyle explanations. The key to this, and to why the conceptualization of 'fatalism' as being opposed to 'lifestylism' is a misapprehension of the real situation, is that while behavioural explanations concern cause, 'fatalistic' explanations deal with distribution.

During the entire fieldwork period, only very few informants were encountered whose belief in randomness led them to deny or ignore the possible benefits of lifestyle change. Rather, the recognition of a pervasive uncertainty in the field of illness and death existed side by side with a common-sense approach to taking appropriate care. Indeed it is the very presence of that recognition that allows lifestyle change to be imbued with any common-sense at all.

Individuals who seem to believe in guarantees of good health through slavish adherence to diets and regimes are widely labelled 'cranks', 'fitness freaks' and 'health nuts'. When a particularly well known 'crank' succumbs to chronic illness (the death from heart failure of Jim Fixx, 'the man who invented jogging' is a good example), the event produces a certain level of public mirth and self-congratulation among the sensible.

Rather than militate against lifestyle change, the recognition of randomness and mystery in the onset of illness and death allows prophylactic behaviour to be managed in a culturally acceptable way. We have observed that an overall guiding principle in popular medical culture is that of balance. Any activity or condition which is recognized as common or normal is seen as safe in moderation, but possibly damaging in excess or absence. A further element in the charting of uncertainty is the widespread recognition that, as each individual is a unique product of heredity, upbringing, exposure to pathogens, personal nature, physical constitution etc, then a belief in the same behavioural norms for all is clearly inappropriate:

> *Informant (Male aged 25 Occ. Class. IIIN)*: '(one person's) body can deal with the intake of food and like and deal with it more efficiently than the other person can ... (Illness can be caused by) maybe uh excess of most things maybe, too much eating, too much drinking ...
>
> *CD*: 'Is there anything doctors can do to keep people healthy? Not necessarily to cure someone if they are ill but to maintain their health, you know to keep them healthy?'
>
> *Informant*: 'Umm I think it's difficult when you actually show an across the board guideline. Umm I think people's needs and attitudes vary so greatly from one person to the next ... I mean certainly if they you know, if they can take into consideration if they sort of push it like so much you know is so much is bad luck.'

Whereas a more general belief that lifestyle change brings certain benefit for all would make it difficult to rationalize 'lapses', an explanatory system that incorporates a certain level of inexplicability allows for patchy or inconstant behaviour. This fits well with a behavioural culture that values 'treats', cyclical indulgence (such as Friday night drinking), the use of recreational drugs as aids to both enjoyment and coping etc. In short, the presence of a developed system of belief in randomness and chance appears to be an aid to the balanced integration into normal life of essentially undesirable behavioural changes (that is, avoiding pleasurable activities and adopting unpleasurable ones).

(A 32 year old Rhondda woman (Occ. Class IIIM) put the case succinctly:

> *Informant*: 'There's a dentist near here. His wife, only a woman in her 30s, was a real fitness fanatic and now she has cancer. To a degree you can determine (your own health), but to a degree it's determined for you. When I hear a thing like that I sometimes think: To hell with tomorrow, why don't you have that cream cake you fancy!'

Conclusions

The research reported here suggests that the personal choice of healthy lifestyles takes place in the context of a sophisticated popular culture. Beliefs concerning the complex aetiology of disease give rise to a pervasive relativism in the perceived benefits of behaviour change. Marked inequalities exist in the relevance of healthy lifestyle advice for different sectors of society. This popular understanding of behavioural risks to long-term good health includes the important recognition that randomness and chance are deeply involved in the intersection of individual lives and general rules. As we have argued elsewhere (Frankel et al 1991) health promotion activities aimed at across-the-board changes in the behaviour of the entire population thus find themselves out of step with popular rationality and the public's understanding of disease causation and distribution.

It follows from this analysis that official and professional attempts to prevent the onset of chronic disease in the population would benefit from a rapprochement with lay culture. Public knowledge and theorizing about the causes and distribution of illnesses is continously informed by developments and changes in the scientific field. A level of consistency in the contents of scientific communication may be something that the health promotion field should concern themselves with. There are few more damaging aspects of the public assessment of messages relating to behavioural health risks than the idea that 'they' are changing 'their' minds yet again. The single most effective way to improve this aspect of communication would be for health promoters and educators to refrain from the use of questionable, unsound or unproven associations in their propaganda. Even where a firm statistical association does exist between behaviour and ill health, it needs to be recognized that the strength of associations vary and that 'lifestyle' propaganda needs to reflect this. It is both dishonest and counter-productive purposefully to promulgate the idea that the proven links between fat consumption and heart disease are of the same order as those between smoking and lung cancer. Even if it were acceptable for public bodies to be involved in the communication of half-truth and distortion (which it is not), the workings of lay epidemiology ensure that 'you can't fool all of the people all of the time' anyway.

While the question of the strength of associations is largely one of standards of honesty in public life, that of the relationship between population statistics and individual lives is more philosophical. Risk associations belong to the world of probability and can readily be described using such phrases as 'one in ten' or 'twice as likely' and so on. The distribution of a 'one in ten' event in a real population, however, is less amenable to concrete descriptive statements. The actual occurrence of individual events in a probabilistic distribution is shot through with uncertainty. The day-to-day activities of health promotion need to address such uncertainty head on.

The fact that Y proportion of obese smokers escape heart disease is as real as the fact that X proportion of the same group succumb. An individual's 'recruitment' to proportion X or proportion Y is an uncertain issue. There seems little point in attempting to escape from or ignore this logical and readily-understood condition.

The background to the public reception of information concerning 'life-style' risks to health is that they exist in an overall landscape of risk. Landscapes are particular to individuals and, as discussed earlier in this paper, include many non-behavioural and non-choice elements. The almost exclusive concentration by health educators and promoters on behaviours said to be open to individual choice, should be re-examined. Mass campaigns aimed at increasing leisure exercise and cutting drug consumption would have more cultural credibility if they were set in the context of public institutions that took issues such as environmental degradation and industrial safety more seriously.

Populations, and groups within populations, have a right to culturally appropriate public services and these should include preventive medicine and health promotion. Combining activities concerning the communication of information about risk with those concerning the social organization of risk distribution was supposedly the rationale behind the absorption of health education into the wider area of health promotion. We have attempted to illustrate here how both functions could address the requirement for cultural legitimacy.

Notes

1 Data are derived from extensive survey and ethnographic research in three communities in South Wales: Plasnewydd (a socially mixed ward of the City of Cardiff), Porth (an industrial and service town in the Rhondda Borough) and Llangammarch Wells (a village and rural district in the Borough of Brecknock). In each area extended semi-structured interviews were carried out with a randomly selected set of adults of both sexes and a full range of socio-economic circumstances (total sample 180). The interviews covered general ideas of what 'health' might be, explanations of the causes of good and poor health, issues of control, prevention, illness avoidance, fault and blame, and a number of more specific topics concerning the heart and IHD. The bulk of these interviews were taped and transcribed. Where interview conditions did not allow the use of a tape recorder, full verbatim notes were made. Field research also involved many hours of observation and participation in local activities, extensive informal contact with residents, and numerous unstructured conversations with key informants. The overall aim of this methodology was, on the one hand, to gather concrete and structured data on the views and experiences of a wide variety of individuals, and, on the other, to attempt a proper contextualization of that data in terms of the daily reality of social life in each

study area. Field research was carried out between April 1988 and December 1989, a total of six months being allocated to each research site.

2 Here we use the term 'healthy lifestyles' to denote those aspects of health-related behaviours and conditions which entail an element of personal action at the individual level. The main fields covered by the rubric are a diet low in saturated fat and refined sugar but high in fibre, the pursuit of leisure activities (especially those involving exercise), and a reduction or cessation of the intake of drugs.

3 In terms of popular nosology, the bio-medical category CHD forms part of the group of ailments known as 'heart trouble'. Not surprisingly, from a social anthropological point of view, the nosological systems of bio-medicine are not entirely co-extensive with those belonging to popular culture, although they have much in common. The conceptual area called 'heart trouble' also includes ailments not immediately definable in terms of the biomedical concept of ischaemia, such as 'weak heart', rheumatic heart disease, 'hole-in-the-heart' etc. Heart trouble due to the blockage or 'furring-up' of arteries (a major cause of ischaemia in the eyes of bio-medical science) is, however, a recognizable sub-category in popular nosology.

4 The 'official' line on personal risk reduction appears in literally thousands of papers, books and pieces of publicity material. For a recent comprehensive résumé of British orthodoxy, see the special May 1987 issue of Health Trends (DHSS 1987). For various contributions to the critique of the dominant paradigm, see Cole (1988), Crawford (1977), Davey Smith, G. (1989), Farrant, W. and Russell, J. (1986), Gillick, M. (1984), Naidoo, J. (1986), and McCormick, J. and Skrabanek, P. (1988).

5 A fairly typical paper which amply illustrates the intellectual poverty of this approach is Lewis et al (1989).

6 A critique of this perspective is provided by Pill and Stott (1987).

7 This is a character trait widely recognized as being inherited via both genetic and social routes. See Davison et al (1989).

This research was funded by the Leverhulme Trust.

References

Blaxter, M. (1979), 'Concepts of causality: lay and medical models', in Oborne, D.J., Gruneberg M.M. and Eiser, J.R. (eds), *Research in Psychology and Medicine*, vol. 2, London: Academic Press.

Blaxter, M. (1990), *Health and Lifestyles*, London: Tavistock/Routledge.

Cole, J. (1988), 'Dietary cholesterol and heart disease: the construction of a medical "fact"', in O'Gorman, J. et al (ed.), *Surveying Social Life – papers in honor of Herbert H. Hyman*, Middletown, Conn: Wesleyan University Press.

Cornwell, J. (1984), *Hard-earned Lives: Accounts of Health and Illness from East London*, London: Tavistock.

Crawford (1977), 'You are dangerous to your health: the ideology and politics of victim blaming', *International Journal of Health Services*, 7 (4): 633–80.

Davey Smith, G. (1989), 'Comprehensive cardiovascular community control programmes', *Abstracts on Hygiene and Communicable Diseases*, **64** (5).
Davison, C. (1989), 'Inheriting heart trouble – the relevance of common-sense ideas to preventive measures', *Health Education Research – theory and practice*, **4** (3): 329–340.
Davison C. et al (1991), 'Lay epidemiology and the prevention paradox – the implications of coronary candidacy for health promotion', *Sociology of Health and Illness*, **13b** (1): 1–19.
Dean, K. (1984), 'Influence of health beliefs on lifestyles: What do we know?', *European Monographs in Health Education Research*, No 6, 127–50, Edinburgh: Scottish Health Education Group.
Department of Health and Social Security (1987), *Health Trends*, **19** (2).
Evans Pritchard, E. (1937), *Witchcraft, Oracles and Magic among the Azande of the Anglo-Egyptian Sudan*, Oxford: Clarendon Press.
Farrant, W. and Russell, J. (1986), 'The politics of health education', Bedford Way Papers 28, London: University of London Institute of Education.
Frankel, S. et al (1991), 'Lay epidemiology and the rationality of responses to health education', *British Journal of General Practice*, **41**: 428–30.
Gifford, S.M. (1986), 'The meaning of lumps: a case study of the ambiguities of risk', in Janes, C.R. et al (eds), *Anthropology and Epidemiology*, The Hague: Reidel.
Gillick, M. (1984), 'Health promotion, jogging and the pursuit of the moral Life', *Journal of Health Politics, Policy and Law*, **9** (3): 369–87.
Lewis, P.A. et al (1989), 'A fatalistic attitude to health among smokers in Cardiff', *Health Education Research – theory and practice*, **4** (3): 361–5.
McCormick, J. and Skrabanek, P. (188), *The Lancet*, 8 October 1988.
Naidoo, J. (1986), 'Limits to individualism', in Watt, A. and Rodmell, S. (eds), *The Politics of Health Education – raising the issues*, London: Routledge and Kegan Paul.
Pill, R. and Stott, N.C.H. (1987), 'The stereotype of "working class fatalism" and the challenge for primary care health promotion', *Health Education Research – theory and practice*, **2** (2): 105–14.
Tones, B.K. (1987), 'Prevention of CHD: the educational task', *Health Trends*, **2** (19): 16–19.
Townsend, P. and Davidson, N. (1982), *Inequalities in Health: The Black Report*, Harmondsworth: Penguin.

8 More medicalizing of mothers: Foetal alcohol syndrome in the USA and related developments

Maureen McNeil and Jacquelyn Litt

Introduction

This chapter is, in part, a case study in the making of a medical syndrome. In 1968, medical scientists first described a syndrome which was definitively named in a *Lancet* article of 1973 as 'Foetal Alcohol Syndrome' (henceforth referred to as FAS) (Jones and Smith 1973). FAS has been linked to a range of symptoms including face and body abnormalities, lower intelligence and abnormal social behaviour – including 'bad judgement'.[1] Since 1979 there has been a regular entry for FAS in the *Index Medicus*, listing articles published in different countries and in different languages.

Although there has been an international response to these developments, researchers in the USA have led the field and our main focus will be on the USA, where FAS has had an increasingly visible public profile.[2] FAS was highlighted in a 1980 report to the President and Congress on health hazards associated with alcohol, in a Surgeon General's Report in 1981 and in a report presented in April 1991 to the President's Committee on Mental Retardation. However, it was not through these, but rather through widely circulated public notices and a flurry of recent media attention that the term – Foetal Alcohol Syndrome – achieved widespread currency in recent years in the USA. Bars, restaurants, liquor stores and beer, wine and liquor bottles sport the following or similar messages:

> *Government warning*: According to the surgeon general, women should not drink alcoholic beverages during pregnancy because of the risk of birth defects.[3]

It is through a recent highly acclaimed book which profiled the syndrome, and through newspaper articles and television programmes that have appeared in the USA in the wake of this publication during the past few years, that FAS has achieved widespread attention.

In 1989 Michael Dorris's book *The Broken Cord* was published, an account of his personal experience with his adopted Native American son Abel (called Adam in the book and henceforth referred to as Adam) who was diagnosed as suffering from FAS.[4] The book became a *New York Times* bestseller, was widely and positively reviewed in the popular press, and has been reprinted several times already. The volume earned Dorris the National Book Critics Circle Award. It has also been advertised in at least one American Women's Studies publication catalogue in the USA (Harper Collins' Women's Studies catalogue:26). The book has triggered other media explorations of this syndrome, most notably, two national television programmes documenting it – a substantial slot on ABC-TV's *20/20*, a PBS special also called *The Broken Cord* and a two-hour television movie with the same title, broadcast on ABC on 3 February 1992. On February 4 1990, the *New York Times Sunday Magazine* ran a feature article on the topic, written by the physician Elisabeth Rosenthal, entitled: 'When A Pregnant Woman Drinks'.

The Broken Cord was the springboard for much of the other media coverage and we see it as a key element in the wave of concern about FAS in the USA. Dorris himself has featured prominently in much of the media coverage. So, we begin with a brief description of the story which the book tells. Michael Dorris is a left-liberal academic who is part native American and was, for some time in the 1970s, a lone parent. The first of his three adopted children was a Native American child – Adam – who began to manifest a series of problems which were eventually diagnosed as FAS. *The Broken Cord* describes Michael's life with Adam and his attempt to grapple with FAS, particularly as it threatens native Americans.

The story is gripping, poignant and well-written. Dorris gains his readers' respect and interest because of his unusual dedication both as a lone parent (particularly a *male* lone parent) and as a fighter for the native American community, but also because he is open about the dilemmas Adam's problems posed. Many readers have no doubt been captivated because it is unusual to find an account of such intense male caring and responsibility for a child.[5] The structure of the book enhances its appeal. The Foreword was written by the acclaimed American writer, Louise Erdich (who is now Dorris's wife) and the last word was given to Adam himself.

The plot has a familiar form – within the genre of the 'cause epic'. Dorris devotes more and more time and energy to investigating the source of Adam's problems which lead him to the diagnosis of FAS. Eventually he gives up his college appointment (at Dartmouth) for full-time research, investigating FAS, particularly in native American communities. The book is designed to raise consciousness about the disease and ends with an invitation for donations to support further research on FAS and related, Foetal Alcohol Effects.

Dorris has clearly earned the admiration of many of his readers, yet we felt uneasy about the book, its popularity and its aftermath. The FAS story as told by him is a moral tale. In this respect, it has much in common with the dominant moral narratives associated with the new reproductive technologies and the more highly publicized example of AIDS. In their analyses of the AIDS discourse, Susan Sontag, Simon Watney and other commentators have reminded us that the packaging of a disease syndrome is rarely abstracted from social and political adjudications (Sontag 1989; Watney 1988). Likewise, we are confronted with what has become 'a moral panic' generated around FAS. In this chapter, we want to raise some questions about that panic and to unpack some of the social and political currents running through and structuring this particular packaging of medical knowledge and popular culture.

Our interest then is in opening a debate where we feel there has been a premature closure. When Barbara Walters (a well-established television commentator in the USA) pronounced that this is 'one of the most important stories' *20/20* had ever covered, we wanted to know how and why such an evaluation was being made. We wanted to understand why Patricia Guthrie (a reporter specializing in Native American alcoholism) who reviewed Dorris's book for the *New York Times Book Review*, found the 'one real, haunting question' to be 'what if someone had stopped Adam's biological mother from drinking?' (Gutherie 1989:20). We also wanted to understand how Dorris could reduce the complexities and ambiguities of the problems in and around FAS to a single enemy – the drinking mother.

The urgency and tone of all these voices indicate that they are dealing with issues that are very important in contemporary USA. In taking as our aim the exploration of this moral narrative, we are aware both that alcohol consumption *can* cause damage in foetal development and that there are problems about alcohol use amongst some women. Yet we felt that it was important to investigate the sub-text that underlies the contemporary discourse around maternal alcoholism and drinking. Dorris's account and others consistent with his book have not precipitated an open exploration of the social and political issues surrounding FAS. Instead, they have narrowed the focus onto blameworthy, guilty parties. In this sense, Dorris offers a relatively clear and conventional solution – improving the behaviour of mothers or punishing them for their deviance.

In order to open debate, we propose to tackle a selection of issues which arise from Dorris's narrative and which relate to a much much broader cultural context. This enables us to situate FAS as the mediator of a complex pattern of social relations.

Big Brother's watching

The villains in this particular moral tale are Adam's mother and other mothers whose drinking is linked to the fate of FAS children. We know this even before we encounter Dorris's own voice, since Louise Erdich is censorious and firm in the Foreword:

> It's a lot of fate to play with for the sake of a moment's relaxation I only hope that she died before she had a chance to produce another child with his problems. I can't help but wish, too, that during pregnancy, if she couldn't be counseled or helped, she had been forced to abstain for those crucial nine months (Dorris 1989:xii).

Erdich mentions situations 'on some American Indian reservations being so desperate that a jail internment during pregnancy has been the only answer in some cases' (Dorris 1989: xvii). Dorris grapples uneasily with the prospect of curtailing human freedom to eliminate FAS. His troubled liberal conscience, however, is overridden by his anger on Adam's behalf and his sense of the magnitude of the threat FAS poses. The finger is clearly pointed at these women. In the absence of consideration of other solutions and generally by allowing others to speak, Dorris seems to endorse the view that

> at some point there have got to be mechanisms in place that are going to make women who are drinking stop having children if they're going to keep drinking (Jeaneen quoted in Dorris 1989:215).[6]

There is a continuity between this call for regulation and the notices in bars and restaurants and on beer, wine and liquor bottles with the state's warning to pregnant women. It would be understandable if some pregnant women hear this as 'big brother is watching you', because, in many respects, big brother has been watching pregnant women more and more vigilantly in the USA and in some other Western industrial countries in recent years. This has occurred against the background of the continuing heated controversy in the USA about abortion legislation and in the wake of continuous challenges to the Roe vs Wade decision which established the legal right to abortion.

Women have been prosecuted (and even imprisoned) for endangering their foetuses for a variety of reasons – for refusing to submit to medical interventions (especially caesarean sections), for using drugs, and for living in ways which are not beneficial to the welfare of their foetuses (vagrancy etc) (Gallagher 1987).[7] While the Supreme Court recently overturned an employer's decision to exclude women from certain jobs in the name of 'foetal protection', it is still in and through state institutions that determinations of women's trustworthiness as mothers is being assessed (Kaminer 1990:21).

Judgements about caesarean sections have been amongst the most controversial manifestations of increased state surveillance of pregnant women. There have been several cases in the USA in which women (some of whom were suffering from serious illnesses or complications) have been forced to have such operations. In 1987 a judge ordered that a caesarean section should be performed against her will on a woman who was dying of cancer. Both the woman and her baby died.[8] As a result of her parents' subsequent appeal on grounds that their daughter had not consented to such surgery and that it violated her right to bodily integrity, the District of Columbia Court of Appeals ruled in a seven to one decision against such action in April 1990. They decided that:

> a pregnant woman, even one who is terminally ill and whose fetus is probably viable, may not be forced against her will to undergo a Caesarean delivery in an effort to save the fetus ('Forced Surgery' 1990:A1).

However, the single dissenting judge formulated the sentiment which has informed this and related interventions when he protested that 'the state's interest in preserving life and the unborn child's interest in life' should be entitled to 'substantial weight' in such decisions ('Ruling' 1990:A11).

Mothers as objects of suspicion are also institutionalized in medical ideology and technology. Pediatric and child-development theories often presuppose maternal inadequacy and dangerousness. Obstetric technological monitoring and genetic testing embody new versions of these suspicions about women in relation to their biological nurturing of the foetus.[9] A recent extension of these assumptions can be found in an article in *Obstetrics and Gynecology* which claims that babies delivered following a difficult and long labour had IQs about ten points lower than siblings who were delivered in scheduled caesarian births ('Study: Hard labor' 1991: 3A).

The responses to and concern about FAS intersect with these developments in identifying the state and medical experts as the guardians of foetal interests which have to be defended against the reproducing woman who is 'under suspicion'. But the state also calls upon others to act as its deputies in overseeing pregnant women. Notices in restaurants and on beer, wine and liquor bottles are precisely such invitations. Dorris refers to two incidents where he and Erdrich 'police' women for drinking during pregnancy (Dorris 1989:199–200). He reveals that 'part of me wanted to make a citizen's arrest'. Barbara Katz Rothman and Katha Pollitt have both reported that pregnant women have been approached by total strangers who reprimand them for drinking (Rothman 1986; Pollitt 1990:409).

Media mothers, foetal voices and women under suspicion

The foregoing activities of the legal system or individual reprimands are not shocking to a public which has become accustomed to representations of women being brought into line with their reproductive roles. In a wave of late 1980s box-office hits, Hollywood presented its own images of potentially unruly women who become 'proper mothers'.[10] *Baby Boom, Immediate Family, Look Who's Talking* and *Parenthood* all revolved around women who in various ways, despite initial deviancy, come to put 'babies first'. When, in *Parenthood* the plot was resolved through every female character of child-bearing age in the film becoming pregnant, viewers are being made more amenable to measures which insist that women are primarily reproductive creatures. When, as in *Look Who's Talking*, audiences come to see a foetus/baby/child as having a will and, through a popular actor, a voice of his own, it is perhaps easier to accept in other contexts that the state or some other designated authority should speak for the foetus *against* the carrying woman.

In many of these conceptualizations and representations the foetus is considered to be a person. Thus, it is not surprising to find Dorris saying:

> One of the hardest things here is to get women to understand that when they drink, the alcohol goes into the placenta, and *the baby's* actually drinking (Dorris 1989:192, our emphasis).

As Rosalind Petchesky has demonstrated, the free-floating, autonomous foetus is an icon of our times (Petchesky 1987). The wide circulation of the image of the autonomous foetus first shown in a series of photographs published in *Look* magazine in 1962 and subsequently in the anti-abortion film, *The Silent Scream*, which appeared in the USA in 1984, has further resonance because of the proliferation of private ultrasound images within many families. These same images were recirculated in *Look Who's Talking* and in the *20/20* documentary on FAS.

The foetus is portrayed in such imagery as 'a little person lying in the womb' (Katz Rothman 1986:115), separated from its mother and this, in part, gives particular resonance to this cultural icon. In addition, an adversarial relation between mother and foetus is structured into this separation and, by now, with the rhetoric around abortion, child abuse and even genetic counselling, we are increasingly accustomed to mother and children being pitted against each other. Juxtaposed to the images of the sentimentalized, nurturant, affectionate mother is her antithesis, the 'anti-mother' or 'monster' who is presented as withholding, cold, uncaring and irresponsible (Tsing 1990). It would appear that every mother embodies this potential 'anti-mother'. Portrayed as innocent and pure, the foetus is seen as being contaminated by the dangerous, monster-like anti-mother. The personification of the foetus and the presumption of its purity and vulnerability make it a powerful symbol in dominant versions of acceptable maternal responsibility.

Maternal responsibilities extended

In many of the developments we have outlined the clinical gaze (in a Foucauldian sense) is central to new forms of regulation of women. However, from a Foucauldian perspective, this is never a matter of merely external imposition: the cultivation of sensibilities, disciplines and knowledges are at the core of such patterns. The history of motherhood can be seen as the elaboration of these sensibilities, disciplines and knowledges. In this particular case, we believe that we have witnessed recently significant extensions of expectations about motherhood into which concerns about FAS play.

The prosecutions associated with maternal responsibilities for foetal development and films like *Look Who's Talking* construct versions of women's responsibilities during pregnancy. They call for further self-surveillance and indicate an intensification of such responsibilities. Faye Ginsberg has argued that the 'crucible' test of female identity and femininity no longer resides in claims to biological/essential femininity, but, rather, in a woman's *'stance toward* her body, and pregnancy in particular' (original emphasis) (Ginsberg 1990:60). The 'choices' offered to more and more women to control their reproduction have rendered traditional categories of biological inevitability unstable. Increasingly, pregnancy and biological motherhood are seen as an 'achievement' (Ginsberg 1990), rather than an ascribed state.

Beyond that, the time-period of maternal responsibility is also being extended. We have by now become familiar with advice about diets and life-style *in preparation for pregnancy.* Recently some pediatricians in the USA have refused to take on a new patient if they have not met the mother when she was pregnant. Furthermore, new reproductive technologies involve an intensification of women's orientation around reproduction in their demanding regimes of clinic visits, examinations, regulation and monitoring of sexual intercourse, recording of body temperature etc, in efforts to become pregnant. Genetic counselling can be seen in this context as part of both patterns of the temporal extension and of the intensification of maternal responsibilities.

In *Parenthood*, there is an amusing incident where the main characters – a middle-class couple in the USA – are shown dealing with a problem child. Summonsed to the principal's office, they are confronted with his complaints. The father turns to the mother and attributes their son's difficulties to the fact that she smoked dope when she was young. The point about this amusing scene is that it clearly represents that pushing back of maternal responsibilities and the expectation that women's lives should be orientated around this destiny from early on. This was clearly and more ominously illustrated in the comments of the Chairman of a United States National Health Report on prenatal care (published in October 1989) who advocated a full programme of adolescent pre-natal education in schools (Kolata 1989).

The debates about FAS are at the centre of this extension and intensification of maternal responsibilities. The syndrome was initially seen as the result of excessive alcohol consumption during pregnancy. But this has since been extended. As a 1983 editorial in *The Lancet* put it:

> Faced with this suggestive but sometimes contradictory evidence the United States Surgeon General and the British Royal College of Psychiatrists have adopted a play-it-safe policy and uncompromisingly advise total abstinence from alcohol during pregnancy (Alcohol and the fetus 1983:683).[11]

This 'zero option' has become more and more dominant as the *New York Times* Sunday Supplement article illustrated (Rosenthal 1990). Because some researchers have argued that conception is a determining period for FAS, Dorris and others have come to advocate that women should not drink at all when they are considering getting pregnant, when they are pregnant or when they are breast feeding. Hence, it is easy to see how the extension of regimes of maternal regulation are legimated and fostered through FAS.

'The modern woman'

While, as we have indicated, suspicion about the mother has a long history in Britain and the USA, this takes on particular contours in the late twentieth century. In a period when motherhood is no longer an automatic feature of women's lives, in which many mothers with young children are employed outside the home, and in which stable definitions of the biological mother are being transformed by technological and judicial developments, maternal drinking has become one arena in which uncertainties about womanhood and motherhood are worked out. Through FAS, dominant values and judgements are circulated about the changing position of women in family units and society. Dorris and others writing about FAS tap into a repertoire of sensitivities about 'the latchkey child', 'the hurried child', the emotionally and physically abused child, the child 'victims' of divorce, and so on.

In *The Broken Cord*, the association between increasing freedom for women and maternal drinking is made quite explicit:

> The ratio of new female drinkers to males has increased by two to one. This disturbing trend (matched by a similar surge in female smoking) may be due to any of several causes, but it undeniably parallels other broad, more positive social changes for women within the same period. However, the process of discarding oppressive and stereotypical roles has not been universally beneficial; liberated of external sanctions discouraging their participation in what were previously guarded as exclusive 'masculine' domains, women are now 'free' to do, and increasingly are doing, just as much harm to their bodies by use of drugs and chemicals as their male counterparts (Dorris 1989:146).

To begin with, we are sceptical about Dorris's vision of social change for women. Who are these women who have been 'liberated of external sanctions', who have benefited from these 'positive social changes'? Was Adam's biological mother, an alcoholic who died from drinking anti-freeze, one of them? Dorris creates a picture of increasing female autonomy which neglects such factors as the continuation of inequality and gender segregation in paid employment, the feminization of poverty and increasing responsibilities for women within the domestic unit.[12] Moreover, he explicitly identifies the 'modern woman' as the one most likely to escape traditional taboos and smoke or drink – as the literal *bearer* of FAS (Dorris 989:226–7).

The assumption of 'modern women's' autonomy combined with Dorris's love for his child leads him to construct the drinking mother as one who wilfully damages her child: 'To some degree, a choice, no matter how buried its trigger, had been made by one party in this sorrowful transaction and was not by the other' (Dorris 1989:193). Much like Nancy Reagan's 'Just Say No' campaign against drugs, Dorris's campaign against FAS (and much other writing on FAS) presumes that alcohol abuse occurs by some positive act of volition that can be corrected by will and choice. Here, as in much of the language of choice used within the abortion debate, liberal traditions and middle-class assumptions are in play. It is presumed that resources to aid in stopping drinking can be obtained and that there are the relevant social opportunities to make the prospect of such change an attractive and realizable goal.

The experience and suffering of the mother and the material conditions that may structure her desperation are neglected. In fact, Adam's mother in *The Broken Cord*, the mothers featured in the *20/20* programme and those discussed in other media coverage are lumped together, as something of a mass of abusing mothers (Tsing 1990; Ginsberg 1990). The FAS discourse, like that associated with 'pro-life' campaigns, flattens individual life circumstances of this kind into the archetypical 'anti-mother'. One of the few attempts to reach out to the mother's experience around FAS was reported on the *20/20* programme when such a mother was interviewed and when the children involved were encouraged to write a letter asking their mothers to explain why they drank. Of course, this is rather double-edged in that it potentially evoked sympathy for the victims and also pointed the finger at *the guilty mother.*

It is certainly and sadly not an original idea to suggest that women's increasing autonomy comes with a high social price. The New Right in both Britain and the USA has issued its own apocalyptic warnings about the end of 'the family' and, concomitantly, the end of civilization. Moreover, behind Dorris's rational presentation of evidence about the increase in female vice lies a lurking suspicion of all female autonomy. It would appear from his account that women's autonomy seems to unleash hedonistic and irresponsible impulses, as if women's responsibility to others – to their children – cease to exist when the social pressures of patriarchy are loosened. In bringing together the image of women under the influence of drink and of women flaunting social conventions, Dorris is superimposing images of women out of control. Whatever else FAS is about, it is primarily about the fears of women being out of control.

Women 'at risk'

There is yet another layer of suspicion that structures the discourse of FAS. This identifies the foreignness or 'otherness' of the Native American woman. In trying to decipher how and why FAS is constructed, we must note that the syndrome achieved attention initially through explicit reference to Native American mothers who drink heavily. According to Sontag (1988), disease is often thought of as being brought by strangers: cholera by immigrants, AIDS by gays and FAS by Native American and poor women. As Anna L. Tsing has put it: 'Even an essential "anti-woman" is created from symbolic scraps of race, class, age, and ethnic difference' (Tsing 1990: 286).

Historically regimes of maternal regulation here involved the implicit or explicit designation of women 'at risk'. The designation of groups of women in any situation as 'at risk' has been a stigmatizing process. For example, Christine Griffin has discussed how the designation of young women as 'at risk' in social work literature involves pathologizations and creates impossible binds for these young women. She has shown that working-class, lesbian and black young women are more likely to be so labelled (Griffin, 1982). Similarly, Valerie Walkerdine and Helen Lucey (1989) have shown the class specificity of the model of 'sensitive mothering' which emerged in post World War II Britain. They show that few working-class mothers could realize this ideal held out. Extrapolating, we could say that, this case study shows how working-class women were designated as 'at risk' in this model of mothering.

The emergence of FAS is clearly about the identification of groups of women as mothers 'at risk'. As the work of Griffin and others suggests, some of us are more 'at risk' than others. Indeed, the term 'at risk', referring to particular pregnant women, does now appear in medical articles about FAS. This is in a context in which the general designation of ever more pregnant women as 'at risk' has legitimated a wide range of medical interventions. The extended use of caesarian sections, ultrasound, foetal monitoring and amniocentesis are but a few examples that illustrate this trend. Hence, FAS is part of this larger pattern of the legitimation of increased medicalization of pregnancies.

In North America it is quite clear that it is the Native Americans (and Canadians) who are under most scrutiny and Dorris's book is very much about them. In surveying the articles listed in the *Index Medicus*, using titles only, it would seem that, except for one study of a specific population in Chile, Native Americans (and Canadians) are the only specific social groups which have been studied in relation to FAS. In a recent report to the President's Committee on Mental Retardation, FAS was related to 'unfavourable socio-economic conditions' ('Poverty as factor' 1991:3a). It is these 'strangers' who are seen as responsible for the circumstances which have threatened white, middle-class adoptive parents and, as Dorris and others would claim, thereby civilization itself.

In search of the perfect child

While some of us have been caught up in debates about postmodernism and disabusing presumptions of progress, parental expectations remain a steadfast bastion of Enlightenment expectations. Children have become the signifiers of the future par excellence in modern Western industrial culture; child development is the definitive paradigm for conceptualizing childhood; the hope that 'our' children will inherit a better world is ubiquitous. But we must be careful with these generalizations. While desires for a better future for one's children might be more or less universal,[13] expectations that this will occur might be difficult to realize if you are a poor parent raising children in impoverished conditions; if you are part of a second or third generation family living in a run-down urban housing project in Britain or the USA or if you are in a poor domestic unit in rural USA where economic and employment prospects are grim.

In Western notions of parental influence, the experience of parenthood can be terrifying, as Michael Dorris eloquently testifies: 'When you decide to have a child, you are hostage to an uncertain future' (Dorris 1989:198). The myth of progress is what sustains this venture and this terror and it would seem likely that this myth has more or less purchase depending on your hold or power in the world. What most parents aspire to is the reassurance that their children have been placed on track to, in some way, '*progress*' in the world. Again, Dorris captures this fundamental aspiration:

'I looked forward to the proud days on which the world would recognize my son as progressively more his own man' (p. 262). Indeed, the constant sense he evokes in *The Broken Cord* is that the problem with Adam is that there is no sense of progress in his trajectory. In a poignant anecdote, Dorris recounts his excitement when Adam gets a job as a caretaker at St Gaudens National Historic Site. At this moment and several others in the narrative, the reader is carried on the wave of Michael's dreams of progress for Adam – maybe this is the change which will get him back on track. This does not happen, and the reader can see the image of 'the broken cord' as a reference to a climbing rope which Michael throws down to Adam and with which he hoped his son could ascend the summit of life.[14]

The Broken Cord and the associated *20/20* television programme record the frustration of parents imbued with this myth of progress with children whose aspirations cannot be realized. As mainly adoptive parents, they are angry that they have not got what they expected, and this is mainly pitted against the biological mothers who are generally from less privileged backgrounds. Indeed, it is important that most of these parents are from the most privileged sectors of society – the white middle-class.[15] This is the group most susceptible to this myth and most inclined to view the mother of their adoped child as 'foreign' or 'other'. But it is also likely that their 'crisis', in part, has been structured by changes in their own historical setting. In a general sense, the faith in progress which sustained this group since at least the Enlightenment has been shaken both theoretically and practically. Meanwhile, their expectations in terms of children have been honed and heightened, helped along by the scientific and 'helping professions' who measure their own successes in terms of child development and progress. It might even be that there is some direct relationship between the diminishing sense of progress in the larger world, and rising investment in child development. Certainly, this is a group which invests extensively in child development. The market has both stimulated and profited from this through the provision of child development manuals, toys, and so on.

Social commentators of various persuasions have noted the intensification of such investment in recent years in the West. They have designated it as the search for, or obsession with, 'the perfect child', or as Barbara Katz Rothman describes it, a tendency to think about children as 'products – the quality of whom can be controlled (Rothman 1986:2, 13). So-called 'new' reproductive technologies and genetic engineering have abetted this intensification, by extending the hope that biological heritage can be more directly controlled by the would-be parents. The campaign around FAS, in this respect, is coterminus with these technologies in so far as it expresses the desires of mainly middle-class parents to exercise more control over that biological heritage – or at least their frustration that they have not been able to control that heritage.

It is important to see FAS contextually as described above. Seen in this way, the anger and, indeed, sometimes rage that is expressed in both *The Broken Cord* and the *20/20* programme comes in part from the fact that these are people who are used to a relatively high degree of public efficacy. Indeed, one of the attractive features of the book is the open way in which Dorris gradually comes to terms with this (although he never locates it as a political insight). Bluntly and assertively, he declares: 'I *willed* Adam to be fine, and he wasn't' (Dorris 1989:71). Only the relatively powerful expect their will to be realized in the world.

We were reminded of Doris Lessing's *The Fifth Child* (1988) which is a novel about the frustrations of a middle-class couple who have a child from whom they feel ever more estranged. In the novel, the mother, Harriet, reflects on the presumptions that have been shattered in this experience: 'We are being punished, that's all ... . For presuming. For thinking we could be happy. Happy because we decided *we* would be' (Lessing 1988: 117). Although in this case the frustration is around biological offspring, there are obvious parallels, and Lessing is presenting an image of the frustration of middle-class senses of entitlement and of the horror of those who expect they can be happy *because they decide they will be.*

This can be contrasted with the limited sense of efficacy many of the native American mothers described in Dorris's book might have had. In this respect, it was perhaps appropriate that Adam was so biblically named, because the myth that sustains Michael Dorris and indeed many middle-class parents (and which theories of child development support and encourage) is very much an origins myth: they can create their child in their own image. The appeal of Dorris's account is his honesty in confronting that aspiration:

> But in the beginning, I didn't think I needed much information: in my naïveté and conceit I was convinced that Adam's life had in a real sense commenced once we were together (Dorris 1989:76).

It is not coincidental that Dorris invokes the metaphor of the child as 'blank slate'. This was the image which inspired Locke and developmentalists since the eighteenth century in the construction of the psychological, social and political frameworks for liberal aspirations.

The horror of Adam's story and of the children shown on the television programmes about FAS is that these are children who in various ways flaunt the conventions of society. The parental nightmare is of an asocial child who espouses no interest in doing what is expected or in fitting in. Adam, for example, far from being challenged by the Solo experience on his Outward Bound trip, sleeps through it. Other children described in these programmes and in the book, sing at inappropriate moments in the classroom, show no fear of strangers, and a host of other symptoms that indicate their lack of social adaptability. For the most part, these children are not productive and thus they can only take from society. This was illustrated in the *20/20* programme which featured one of these FAS 'victims' in his prison cell. Thus these individuals are repeatedly presented as burdens on society.

Expectations about choice and control have been heightened for contemporary middle-class parents or would-be parents in North America and Britain. In a world of new reproductive technologies, some of them are trying to shape the biological constitution of their children.[16] The reorientation of modern medicine around genetic engineering encourages such expectations. The high quality products promised by agencies like the Nobel sperm banks promise selectivity – the new mail order catalogues of the middle-class USA. Moreover, it has become more common for Americans to regard the whole world as potential shopping ground in their search for their 'own children', as international adoptions become more common. With scares about high rates of infertility (to which the middle-class are particularly susceptible), it is not surprising to find middle-class parents turning their attention to the pool of adoptive children available and to be concerned with improving that biological stock or their choices in that field as well.

Apocalypse now

The Broken Cord is a book about collective disaster, as well as personal tragedy. By linking the crime rate, alcohol and drug abuse, illegitimacy and neglectful mothers to the problem of FAS, Dorris pushes beyond the sphere of the personal onto a broader social panorama. There is an apocalyptic tone to his account and in some of the associated media coverage. Maternal drinking could spell the end to the Native American community. Dorris claims that a quarter of all Native American children suffer from the infliction in some form. That these communities have survived the 'invasion and plague and government policies aimed at cultural genocide' may be a cruel irony, as Dorris contemplates that 'our story [might] end with alcohol, a liquid so deceptively fragile that it could evaporate in ordinary air' (Dorris 1989: 183). Having survived external threat, the Native American community could now self-destruct, and the drinking mother has been identified as the agent of that apocalypse (Stange 1990).

While FAS is a particular threat to the Native American community, Dorris sees this prophecy of biological apocalypse extending further, because FAS 'knew no ethnic boundaries' (Dorris 1989:226). He quotes Dr Phillip May's observation: 'If we have a lot of drinking mothers, we may someday have whole societies with lower average intelligence' (Dorris 1989:183). These fears are further exacerbated when Dorris and others consider that rates of women's drinking are increasing and that foetal alcohol victims are more likely themselves to give birth to foetal alcohol victims. These prospects for the reproduction and spread of the problem give it an epic quality.

Reading Dorris's account, it is difficult not to be reminded of the eugenic fears and their proposed solutions from the early twentieth century. Dorris is concerned about both the quantity of children that these mothers produce, as well as their quality. These are people who cannot be productive and who will be costly to the middle-class both personally in individual adoptions and, more broadly, socially and economically, in their states of social dependence.

Dorris's prophecy of biological destruction echoes much of the rhetoric surrounding AIDS. In both cases, the disease phenomena is linked to the possible end of civilization and both are seen as 'inflicted' in the sense that they are considered to be the consequences of some larger moral degeneration (Sontag 1988). The fact that neither AIDS nor FAS are curable makes them particularly amenable to their overburdening with social significance: implicit but salient dominant cultural values come to the fore in the search for solutions. Hence, FAS, like AIDS, is often presented as a social drama, equivalent to the 'medieval morality play' in which late twentieth century dominant concerns about social progress and middle-class efficacy, children, maternity and women are acted out.

The analogy between AIDS and FAS could be pushed yet further since both represent in some ways graphic examples of the limits of scientific progress. They are perhaps equivalents in the biological realm to the Challenger disaster which was quite literally the explosion of an important public symbol of scientific and technological progress. The crisis creates related fears: Is modern science and medicine still unable to eliminate the spector of epidemics? FAS seems to involve a group of people who are 'not whole', whose destiny is 'fixed in utero and modified only slightly and perhaps temporarily by anything that happens after birth' (Dorris 1989: 149). The limitations of contemporary science and technology and of the American dream of meliorism, activism and control are all at stake in these three public dramas focused on science and technology.

It may well be that this crisis of faith in the progress of science and technology takes on special resonance because it is seen to be a crisis about mothering, involving class and ethnic conflict. The dilemmas surrounding FAS involve a case in which a biological mother's pollution is seen to threaten not only the pure, innocent foetus, but also the parental expectations and dreams of white, middle-class Americans.

Conclusion

This has been a beginning in the analysis of how FAS (in the USA, in particular) serves as a drama in which dominant ideas about motherhood, social progress and a range of other issues can be played out. We have already witnessed ways in which the 'problem' of maternal drinking associated with FAS has seeped outside the domain of medical journal reports and into television programmes, newspaper features and the committed autobiographical best-seller of Michael Dorris's *The Broken Cord*. The *New York Times* Sunday supplement article, written by an established medical authority, undoubtedly conjured fears amongst many middle-class Americans, with its account of a seemingly endless range of symptoms including virtually any form of socially abnormal behaviour and its warning that *any* drinking at the time of conception, during pregnancy or during breast feeding was dangerous. Moreover, complementing the labels on beer, wine and liquor bottles and notices in American restaurants and bars, prime-time television has added its voice to the chorus in the popular *Thirty-Something* series where the pregnant character – Hope (appropriately named, perhaps?) – was shown refusing alcohol.

It seems that feminists are also being recruited to this campaign. As mentioned above, *The Broken Cord* has appeared in at least one Women's Studies catalogue under the listing for 'Health' (Harper Collins 1990). *The New Our Bodies, Our Selves* (1991) carries the following comment: 'As little as two drinks per day during pregnancy has been associated with lesser effects, notably lower birth weight in newborns. At present, no safe level of alcohol use during pregnancy has been established.' It continues in a note: 'Because the federal government has withdrawn plans to place warning labels on alcoholic beverages, public interest groups need to apply pressure to revive these plans' (Sandmaier 1984:33). This could be considered to be another attempt to educate women about their bodies, which is the main agenda of *Our Bodies, Our Selves*. Yet, as we have indicated, there is more at stake in this issue. We need to examine what it means to use public warnings as a means of educating women. Such advocacy presumes, as does Dorris, that maternal drinking is a matter of will. Moreover, such warnings can justify and animate vigilantism in social interaction and in state control of women in the name of foetal rights (Pollitt 1990). Perhaps even more insidiously, such endorsement ignores the fact that in this and in the increasing number of judicial cases against women, the state's concern is not motivated by women's best interests. This would also suggest that we should be wary about new pre-natal programmes which can be used as justifications for regulation and monitoring (particularly for groups designated as 'at risk' and thus, blameworthy). As much of our paper has demonstrated, the prevailing climate is one of suspicion towards women (pregnant and otherwise) and there is less concern about what happens to them, their health, etc. than there is for the foetus and the child.

The history of pre-natal care in both the USA and Britain indicates a need for caution. In the USA during the Progressive Era, when child health occupied a central place on the national agenda, 'mothers (particularly working-class) made a useful scapegoat' in references to infant sickness and death (Davin 1978). Poverty, poor sewage, bad living conditions, the toil of domestic labour and paid employment were not considered as so significant in analyses of the causes of infant mortality. While a number of initiatives from Mothers' Clubs to dispensaries were launched to educate mothers, it was not until the infant mortality rate began to decline that maternal mortality became an issue for public concern. With this historical background we can see that campaigns around maternal and infant health have generally carried a set of 'national interests' – productivity, imperialism, and so on. Moreover, they have often been defined not by mothers themselves or necessarily in their interests (Lewis 1990; Davin 1978).

The controversy surrounding FAS needs to be seen in its full context. Much of the discussion of FAS referred to in this paper obscures the conditions that produce alcoholism and diverts attention away from the fundamental inequalities in social structure that affect women's health. For this reason, we feel it is necessary to turn attention towards the issues that are *not* addressed in an exclusive focus on the FAS mother. First, medical scientists concentrate much of their research efforts to the investigation of female, rather than male reproduction. Although more attention has been given recently to work environments which may adversely affect the reproductive capacity of men, this is a relatively recent development and not insignificantly, this research focuses on environmental issues more than on individual, personal responsibility. Second, male alcoholism and its impact on children (and indeed, on the health of pregnant women through violence and other maltreatment) has not come under such scrutiny. Indeed, Dorris rather glibly observes that: 'Drunk expecting fathers may, at least as far as it is currently known, hurt only themselves and those unfortunate enough to get in their way' (Dorris, 1989:146).[17] Third, at a time when these communities are bringing more attention to the racism and disinheritance that has shaped their living conditions, FAS constructs a terrifying story of self-destruction and points the finger back at the community itself – particularly its women. Dorris himself recognizes this with his troubled asides, as he reflects on the implications of where his argument might lead: 'Forced sterilization, paternalistic government programs, men making decisions about women's bodies – the stuff of federal Indian policy for the past hundred years' (Dorris 1989:186). The final neglected element in the dominant versions of the FAS story is the story of cuts in welfare provision (what might be described as the decline in the assumption of collective responsibilities) which provide the back-drop for concerns about FAS and other alleged instances of maternal personal irresponsibility (Platt 1991).

The danger is that it is easy to get locked into the dominant discourse, which denies these complexities. The medical model mobilized by Dorris and circulated in the media and some popular culture has primarily been orientated around maternal blame. A feminist reconceptualization would begin not by warning women of the dangers of drinking, but by rethinking the problem of maternal drinking. This requires placing women back at the centre of the analysis – not as the agents of destruction – but as persons with experiences, feelings and needs. Beyond this, FAS itself is at the centre of, and to some extent has come to stand for, a complex pattern of relationships, desires and expectations. We would not deny that it is about maternal drinking, but like so much medical knowledge, it is about so much more.

Notes

1 It is interesting to look at definitions of FAS. In an article in the *New York Times* Sunday Supplement, Dr Elisabeth Rosenthal offered the following definition:

> Fetal alcohol syndrome and its more subtle variant, fetal alcohol effect, are umbrella terms used to describe the condition affecting the scarred offspring of drinking mothers. Victims with full-blown syndrome, whose mothers generally drink heavily throughout pregnancy, often suffer physical malformations and mental retardation. Even those less fully affected, sometimes the progeny of women who drank only intermittently, may end up with lifelong learning disabilities and behavioural problems (Rosenthal 1990:30).

It should be noted how many references to the 'guilty' mother appear in this short definition. Aside from this issue, a full study of the range of symptoms attributed to the syndrome would itself be very revealing.

2 There has been some media coverage of this issue in Britain, particularly since 1990. Generally, there has been less panic about in Britain and more equivocation in the media coverage of the issue of women drinking during pregnancy.

3 From November 1989 all manufacturers of beer, wine and liquor in the USA were required to put labels on their products warning of the association between birth defects, health problems and alcohol consumption. From April 1991 New York and nine other states required every bar and restaurant in their states to post a warning about the dangers of drinking alcohol during pregnancy. The New York branch of NOW (the National Organisation of Women) opposed this legislation.

4 Abel Dorris died in September 1991 from injuries sustained in a car accident.

5 It is interesting to note that television situation comedies in the USA have

been increasingly featuring the 'motherless' family. Fathers are often presented in these programmes as bumbling and relatively incompetent, and the children as suffering in some ways. Such programmes are likely to evoke sympathetic feelings towards the father. For a full and critical evaluation of recent concerns around fathering in Australia, Britain and the USA, see Segal 1990:27–59.

6 The idea that state policy should set limits on women's reproduction is no longer hypothetical in the USA. Recently, a seventeen year old woman was sentenced to two years in prison and was required to use birth control for ten years following her release for allegedly 'smothering her newborn daughter in the bathroom of a ... hospital' ('Sentence' 1990:A1).

7 These prosecutions are often conducted under the rubric of 'child abuse' or 'child neglect'. Such rubric contributes to the construction of the distinct interests of the foetus and of foetus 'personhood' with implications that have been discussed by Rosalind Petchesky and others. See: Petchesky (1987), Tsing (1990), Katz Rothman (1986). Such constructions have been important in the controversies surrounding abortion and some aspects of their significance are discussed below. See also Pollitt (1990). There have now been a number of charges of abuse, assault and child endangerment brought against women in the USA in relation to their use of alcohol. For an account of similar cases see Tsing (1990).

8 This incident provided the basis for an episode of *LA Law*. In that version, the baby lived. See Pollitt, (1990).

9 See Arney (1985) for an interesting account of how historically obstetric theory conceptualized this suspicion.

10 *Grifters* is another recent Hollywood film which should be mentioned here for its extremely negative portrayal of mothers.

11 The reference to the ambiguity of the evidence should be noted. Moreover, the editorial draws support from an article in the same issue of the *Lancet* by Wright et al (1983). However, this article does not mention the 'zero option'.

12 It should be noted that British commentators have also brought attention to patterns of increasing smoking and drinking among women.

13 See Dill (1980) for a discussion of the desires of poor African-American mothers.

14 The more obvious analogy in play here is to the umbilical cord. The title of the Dorris book can be read in a variety of ways. It is likely that these readings would revolve around the severing of biological links between the biological mother and child. One reading would be that the title underscores the responsibility of those women who are acting 'against nature' and severing positive natural connections to their future child through drinking during pregnancy. Another possible reading is that Dorris is actually trying to severe the connection between Native American mothers and their offspring (see Stange 1990).

15 This is not true of Dorris himself, although he enters fully into this belief in progress. Indeed, we would suggest that part of the appeal of the book is the fact that it is written by a Native American who fully enters into these expectations which are most central to the values of white, middle-class America.

16 For a discussion of class and racial difference in regard to genetic testing see Rapp (1990).

17 Obviously much depends on how you interpret 'unfortunate enough to get in their way'.

References

'Alcohol and the fetus – is zero the only option' (editorial) (1983), *The Lancet*, **i**, 26 March, pp 682–3.

Arney, W.R. (1985), *Power and the Profession of Obstetrics*, Chicago: University of Chicago Press.

Davin, A. (1978), 'Imperialism and motherhood', *History Workshop Journal*, **5**:9–65.

Dill, B.T. (1980), 'The means to put my children through: child-rearing goals and strategies among black female domestic servants', in Rodgers-Rose, La Frances (ed.), *The Black Woman*, Beverly Hills: Sage.

Dorris, M. (1989), *The Broken Cord*, with Foreword by Louise Erdrich, New York: Harper and Row.

'Forced surgery to save fetus is rejected by court in capital', (1990), *New York Times*, 27 April 1990, pp A1, A11.

Gallagher, J. (1987), 'Prenatal invasions and interventions: What's wrong with foetal rights', *Harvard Women's Law Journal*, **10**:9–58.

Ginsberg, F. (1990), 'The "Word-Made" flesh: the disembodiment of gender in the abortion debate', in Ginsberg, F. and Tsing, A.L. (eds), *Uncertain Terms: Negotiating Gender in American Culture*, Boston: Beacon Books.

Griffin, C. (1982), 'The good, the bad and the ugly: images of young women in the labour market', Birmingham: Centre for Contemporary Cultural Studies Stencilled Paper, No. 70.

Guthrie, P. (1989) 'Alcohol's child: a father tells his tale' (review of the *Broken Cord* by Michael Dorris), *New York Review of Books*, 21 August, pp 1–20.

Jones, K.L. and Smith, D.W. (1973), 'Recognition of the Foetal Alcohol Syndrome in early infancy', *The Lancet*, **ii**: 999–1001.

Kaminer, W. (1990), 'The foetal protection charade', *New York Times*, 29 April, p 21.

Kolata, G. (1989), 'Less prenatal care urged for most healthy women', *New York Times* (4 October), p 1.

Lessing, D. (1988), *The Fifth Child*, New York: Random House.

Lewis, J. (1990), '"Motherhood issues" in the later nineteenth and twentieth centuries', in Arnup, K., Levesque, A. and Pierson, R.R. with the assistance of Brennan, M. (eds), *Delivering Motherhood: Maternal Ideologies and Practices in the 19th and 20th Centuries*, London: Routledge.

Petchesky, R.P. (1987), 'Foetal images: the power of visual culture in the politics of reproduction', in Stanworth, M. (ed.), *Reproductive Technologies: Gender. Motherhood and Medicine*, Cambridge: Polity Press.

Platt, S. (1991), 'Fertility control', *New Statesman and Society*, 28 June, p 11.

Pollitt, K. (1990), 'A new assault on feminism', *The Nation*, 26 March, pp 409–18.

'Poverty as factor in retardation, study says' (1991), *The Philadelphia Inquirer*, 26 April, p 3a.
Rapp, R. (1990), 'Constructing amniocentesis: maternal and medical discourses', in Ginsberg, F. and Tsing, A.L. (eds), *Uncertain Terms: Negotiating Gender in American Culture*, Boston: Beacon Books.
Rosenthal, E. (1990), 'When a pregnant woman drinks', *New York Times Magazine*, 4 February, pp 30, 49, 61.
Rothman, B.K. (1986), *The Tentative Pregnancy: Prenatal Diagnosis and the Future of Motherhood*, New York: Viking Penguin Inc.
'Ruling by capital court on foetal status' (1990), *New York Times*, 27 April, p A11.
Sandmaier, M. (1984), 'Alcohol, mood-altering drugs and smoking', in The Boston Women's Health Collective (ed.), *The New Our Bodies. Ourselves: A Book by and for Women*, New York: Simon and Schuster Inc., pp 33–40.
Segal, L. (1990), *Slow Motion: Changing Masculinities. Changing Men*, London: Virago.
'Sentence for killing newborn: jail term, then birth control', *New York Times*, 18 November, p A1.
Sontag, S. (1989), 'Aids and its metaphors', *New York Review of Books*, October, pp 89–99.
Stange, M. (1990), 'The broken self: Fetal Alcohol Syndrome and American Indian selfhood', paper presented to the American Studies Association Meeting, November.
'Study: hard labor may cut baby's IQ' (1991), *Philadelphia Inquirer*, 1 May, p 3A.
Tsing, A.L. (1990), 'Monster stories: women charged with perinatal endangerment', in Ginsberg, F. and Tsing, A.L. (eds), *Uncertain Terms: Negotiating Gender in American Culture*, Boston: Beacon Books.
'Two dismissed in warning on alcohol and pregnancy' (1991), *New York Times*, 30 March, p 7a.
Vieazo, F. (ed.) (1991), *The New Our Bodies, Our Selves*, New York: Harper Collins.
Walkerdine, V. and Lucey, H. (1989), *Democracy in the Kitchen*, London: Virago.
Watney, S. (1990), *Policing Desire: Pornography, AIDS and the Media*, 2nd edn, Minnesota: University of Minnesota Press.
Wright, J.T., Barrison, I.G., Lewis, I.G., MacRae, K.D., Waterson, E.J., Toplis, P.J., Gordon, M.G., Morssi, N.F. and Murray-Lyon, I.M. (1983), 'Alcohol consumption, pregnancy, and low birthweight', *The Lancet*, i, 26 March, pp 663–7.

9 'What's your excuse for relapsing?': A critique of recent sexual behaviour studies of gay men

Graham Hart, Ray Fitzpatrick, Jill Dawson, John McLean and Mary Boulton

Most national and local governments have directed specific monies to AIDS research, services and treatment. For social scientists involved in behavioural studies of populations at risk of HIV infection – deemed necessary for the purposes of epidemiological surveillance and health education – resources have been made available to undertake large surveys, small qualitative studies and studies that evaluate the efficacy of particular behavioural and health interventions. In particular, a massive amount of data has been generated in the developed world on homosexual and bisexual men. From a sociological viewpoint, however, the production of such data has not been interpreted by reference to any particular theoretical perspective or related to the now considerable body of knowledge in the sociology of health and illness.

The aim of this chapter is to attempt a sociological critique of a concept recently introduced to behavioural research on gay men, and to do this by reference both to the empirical and epistemological bases of what at present is a mainly North American literature. We will begin by summarizing the epidemiological and behavioural data which established the extent of the adoption of HIV risk reduction strategies by gay men, and then describe the introduction of the concept of 'relapse' to account for observed trends towards so-called unsafe sexual behaviour. The use of this term will be subjected to an empirical critique, supported by data from the UK; we follow this with a conceptual analysis of the episteme of male sexual behaviour employed in these studies. Finally, the paper will end by suggesting why the term 'relapse' should appear at this moment in the AIDS epidemic in the USA, but also the direction that a more sociological analysis of behaviour might take.

Evidence of behavioural change

It is now well established from studies of homosexual and bisexual men in the USA, Europe and Australia that there has been a major move towards the adoption of 'safer sex' behaviours in this population (Hart 1989). Evidence for this comes from three sources. Firstly there has been a reduction in all sexually transmitted diseases in gay men since the early years of the decade, indicating a general reduction in risk for infections which have similar transmission characteristics to HIV (Johnson and Gill 1989). Secondly, in the UK since 1986 there has been no increase in the level of HIV in men attending genito urinary medicine clinics, so avoiding the rapid incidence of infection seen in the early days of the epidemic in San Francisco or New York (Loveday et al 1989; Evans et al 1989). Finally, interview and questionnaire-based studies relying on self-reports of behaviour change have consistently indicated a reduction in the primary risk behaviour for HIV infection in gay men (unprotected receptive anal intercourse), increasing use of condoms in penetrative sex, and a fall in the numbers of sexual partners reported (Fitzpatrick, Boulton and Hart 1989).

However, it rapidly became clear that, although on a population basis there had been behaviour change in the direction of safer sex, not all participants in the self-report studies had adopted sexual behaviours which were free from risk of HIV infection. From the epidemiological and health education perspectives the question that now arose was what characterizes those men who have unsafe sex?

Explanations of behaviour change

The studies which sought to answer this question have been described in a review which gives an account of the demographic and psycho-social variables that distinguish those having unprotected anal intercourse (synonymous in this literature with 'unsafe sex') from those who do not (Hart 1989). Research has identified demographic, cultural and socio-economic variables associated with differences in reported behaviour, but has primarily focused on psycho-social factors, particularly attitudes, associated with behavioural change.

Many of these studies have been guided by the Health Belief Model (Rosenstock 1974a; 1974b) which suggests that people respond to health messages on the basis of the perceived severity of disease, risk to themselves and the costs and benefits of changing their behaviour (Fitzpatrick, Boulton and Hart 1989). This model, and the even more psychologistic Health Locus of Control model which emphasizes personal efficacy and fatalism in response to health problems (Wallston and Wallston 1981), can be criticized for ignoring the economic and social bases of human action (including or more particularly sexual interactions) and for viewing individuals as free to make choices only on the basis of carefully considered and rational thought, with no other consideration of social structural location. For example, it is clear from the work of feminist writers and researchers that when one considers gendered structures of inequality and the exercise of power it may not be possible for many women to follow recommended guidelines regarding condom use, even when this is desired (Richardson 1990; Wilton and Aggleton 1991; Holland et al 1991). However, imbalances of power are not the only consideration; individuals who have the power to control their behaviour will not always adopt changes prescribed by health professionals. There may also be situational factors associated with action – such as the availability of condoms at the time sex takes place, or emotive/affective responses to a situation which culminate in the suspension of recommended behaviour.

The aim of the Health Belief studies has been to provide health educators with the necessary information to design interventions suited to the needs of individuals in the 'unsafe' category, to encourage them also to change their behaviour. As longitudinal studies, that have enrolled cohorts of gay men followed over a period of months or years, they have not only investigated behaviour change, but the *maintenance* of such change. It soon became clear that, whilst a majority of men reported only 'safe sex', either consistently throughout a study period, or starting during it, another group – initially reporting only safe sex – said they had had unprotected anal intercourse during the follow-up waves of the studies. It is this group, and the terms used to describe the men involved, that we will now address.

'Relapse' to unsafe sex

The 1989 International Conference on AIDS was held in Montreal, Canada, and in 1990 in San Francisco, California. The abstract books from the Montreal conference included only one poster using the term relapse (Ekstrand, Stall, Coates and McKusick 1989), whereas the 1990 conference saw its widespread use in the analysis of data sets. 'Relapse' was used to describe the behaviour of that minority of men who had initially adopted safer sex behaviours but, at some point during follow-up, had had at least one episode of unprotected penetrative anal intercourse. It appeared that a 'new' category of men had been uncovered – those who revert back to their former sexual ways in the later stages of the epidemic. By 1992 'relapse' had been incorporated into the health education vocabulary of prevention, and posters in San Francisco challenged gay men to report their 'excuses' for relapse to unsafe sex to a health counsellor at the end of a given telephone line. What is the evidence for the existence of this group?

One of the largest studies to monitor gay sexual behaviour is the AIDS Behavioral Research Project, a San Francisco based study very influential in health education programmes. This reported that 71 per cent (n=575) of a cohort initially recruited in 1984 were still in the study in 1988 and, of these, 397 had responded at every wave of measurement and had been sexually active during at least one of the five waves of measurement. Whilst only 2.3 per cent of the men reported at least one episode of unprotected anal intercourse at each of the five waves, 19 per cent 'reported relapse after having only safe sex at a previous wave' (Stall et al 1990a). In fact, respondents were not asked to 'report relapse', simply to describe behaviour in the previous month (thirty days) at each wave prior to 1988, and the previous month and year in 1988. Using variables which had been employed in 1984 that had a predictive value in relation to likelihood of engaging in unsafe sex, it was found that identifying unprotected anal sex as a favourite act and having 'social support for taking health-related risks' were significantly associated with 'relapse' amongst the forty six men who had initially reported safe, and then later reported unsafe, sexual behaviour.

However, looking at a more detailed report of the 1988 data (Stall et al 1991b) for the last month and year, it became clear that, of all men, including 'relapsers', reporting unprotected anal intercourse (256 of 535) the largest single group were in mutually monogamous relationships (47 per cent), 18 per cent were in regular relationships that were not mutually monogamous, and 35 per cent had no primary relationship. Taking an overview of the whole sample, just over half of all of those in mutually monogamous relationships reported at least one episode of 'unsafe sex', whereas 36.7 per cent of the non-monogamous men and 36.4 per cent of the men without a primary relationship reported such an episode. Asked to either agree or disagree with closed-ended questions as to why this had taken place *or* to write in their reason for this, men in monogamous relationships were more likely to report having the same HIV status as their partner (48 per cent) or being in love with their partner (41.4 per cent). Men in non-monogamous relationships and those with no primary relationship were more likely to identify being 'turned on' (59.6 per cent; 70 per cent) as their reason for unsafe sex, although in these two groups knowing the antibody status of one's partner was reported frequently (42.6 per cent; 21.1 per cent), as was knowing that one's partner was HIV negative (monogamous, 13.8 per cent; non-monogamous, 23.4 per cent; no primary relationship, 13.3 per cent). Stall et al suggest that their primary goal in presenting these data 'has been to stimulate thinking about the prevention of relapse from safer sex techniques' (ibid:1186).

Data analysed by members of the same research team of men recruited to a different cohort and over a complete four year period also supported the relapse thesis (Ekstrand and Coates 1990); this was at the level of 12 per cent. Four other research projects reported at the 1990 Conference used the term 'relapse', and all were from the United States (Bartholow et al 1990; O'Reilly et al 1990; Adib et al 1990; St Lawrence et al 1990). Finally, the term 'prevention failures' was used by one group of Canadian researchers to describe men who, when first recruited to a longitudinal study in 1985, had been HIV-antibody negative but by 1989 had seroconverted (Willoughby et al 1990). These authors, although using a highly mechanistic concept, employed a clear measure of true level of risk by showing the outcome of such risk-taking, namely infection with HIV. None of the other studies applied this criterion, and were concerned only with risk behaviours. This could be one criticism of these studies, but in the following sections other methodological and conceptual bases for a critique of concepts such as relapse and prevention failure will be explored.

'Relapse': empirical critique

The first point to make is that relapse is *not* a term employed by study participants. Interviewees or survey participants were not asked why they had 'relapsed' rather why they had a certain type of sex in a specific context or with particular partners. Unlike the term 'safer sex', which it has been suggested was invented by the gay community (Paton 1989) and has certainly been adopted widely, relapse comes from a medico-psychological perspective not necessarily shared by men having gay sex.

It should also be noted that the absolute numbers of men in these studies 'relapsing' to unsafe behaviour is small; 19 per cent in Stall et al (1990b) equalled seventy-four men, and 40 per cent in St Lawrence et al (1990) represented twenty-seven individuals. The majority of men in all the studies report maintaining safer sex practices. Indeed, in Stall et al the proportion of men who had 'changed to lower risk' was 30 per cent (n=119), yet this was not considered to be of sufficient note to comment upon because of the relapse emphasis in the paper. Even on their own terms many studies could be underestimating the amount of unprotected anal intercourse that occurs because of the unrecorded behaviour of men lost to follow-up, the use of any period which does not include all the time between study waves and by recruiting mainly white middle-class gay men with a clear commitment to stay with the studies over time, and whose behaviour has seen the greatest degree of change.

However, it is the next point that is perhaps the most important. That is, the designation of the men's sexual behaviour as 'safe' or 'unsafe'. Stall et al (1990b) distinguish between those at 'modified high risk' (men having unprotected anal sex within a monogamous relationship) and those at 'high risk' (anal sex without condoms outside monogamous relationships) and say that '... from a prevention point of view any form of unprotected anal sex is undesirable' (ibid:1182). This expresses most succinctly the view that, from a health education perspective, all penetrative acts should be protected, regardless of circumstance. Yet it is clear that not all forms of unprotected anal sex carry risk for HIV infection. This is most apparent for those men who are in monogamous relationships and who are aware of their own and their partner's HIV antibody status. In the absence of other infections or ill health, there is no reason why these men should not have 'unsafe' sex.

If a definition of safe sex is used which is restrictive, then methodologically the extent of so-called unsafe sexual activity will be exaggerated, and the means used by people either to avoid infection, or to live with and accept the risks associated with a particular activity, will be ignored. For example in some reports (Adib et al 1990; Kelly, St Lawrence and Brasfield 1991) partner type is not specified, and relapse becomes an absolute category regardless of the nature of relationships or knowledge of antibody status. In a study also presented at the San Francisco Conference but which did not employ the term relapse (despite giving an account of a cohort) it was made clear that

> men engaging in unprotected anal receptive sex were more likely to be in long-term primary relationships, to have lovers with known negative serostatus, and to be in relationships closed by agreement to outside sex. Men in open relationships or men who suspected their partners were unfaithful were more likely to use condoms. ... (Ames and Beeker 1990).

What we are not suggesting is that there should be a new health education programme proposing careful partner selection, or that monogamy as such is protective against HIV. Clearly to rely on trust as a means of protection is insufficient as a barrier to infection. However, it is the case that as the epidemic and responses to it change, varying strategies for living with the presence of a life-threatening disease develop. There could be a recognition of this, whilst continuing to encourage safer sex in any situation of possible infection.

Even when men do not know their antibody status or that of their partners, and have unprotected anal intercourse because of love, trust or possibly coercion within a relationship, they are likely to be aware of the consequences of their actions. Most research reports no longer describe levels of knowledge about HIV transmission in study populations as these have reached very high levels; participants know in particular of the risks attendant on anal intercourse, and many have a sophisticated knowledge of treatments as a result of articles in the gay press. Men are taking a risk on the basis that other information about their partners – such as sexual history and possible exposure to HIV – is either sufficient for them to have unprotected sex or, and just as important, *that they are willing to run the risk of infection*. Whether one employs the Health Belief Model to explain this, or a socio-structural or interpersonal model which accounts for imbalances of power, these men have either made a decision to engage in a particular type of sexual behaviour, and thus accepted the risks involved, or have implicitly consented to its taking place.

Even if one accepts that 'relapse' has occurred, its application to some men is inappropriate. Men initially having 'safe sex' – or doing so when studies began – may have gone on to have unprotected anal intercourse for the first time during a study period. This could apply to some of the younger men in studies, or to those who had simply never had anal intercourse as part of their sexual repertoire. Thus, these men are not *relapsing* to unsafe sex but are *commencing* to have it.

A methodological issue recognized by some of the researchers involved in studies of 'relapse' is that, by applying to 'relapsers' the demographic, behavioural, social and psychological variables used initially to distinguish those who practise safe from unsafe sex, researchers are once again identifying men who have 'unsafe' sex. That is, the 'relapsers' are in psychosocial terms indistinguishable from the men who have been consistent in having unprotected penetrative anal intercourse; they happened to be having 'safe sex' at the outset of a study, for example by having no partners or (temporarily) no penetrative sex, but later returned to their usual behaviour pattern. This group would also include men who in behavioural terms were inconsistent in their condom use, and had been recruited to the study at a time when they were temporarily using condoms. Stall et al relate this to the limitations of using a thirty day period for sexual behaviour.

> The use of this measurement time frame ... allows for a possible misclassification of each of the risk categories. For example, some of the 'relapsers' could perhaps be more accurately described as chronic high risk takers if all sexual behaviour over the entire 5 years was considered.

We would go further than this and suggest that the group described as 'relapsers' may be coterminous with men who are either inconsistent in their use of safe sex *or* are highly selective in the kind of sex they have with particular partners, as a protective strategy. They may be fully cognisant of the risks associated with their behaviour, or at least suspect a degree of risk, and in their own terms be acting on the basis of an informed decision about the nature of sexual behaviour in the context of a specific relationship.

The application of 'relapse' to UK data

In a sense these methodological issues arise purely from a reading of the present literature, but another way of approaching the issue of relapse is to apply the concept to data which has been collected in a similar way. Since

1987 we have been undertaking a longitudinal research project on gay and bisexual men in the UK and have published data on the demographic and social characteristics of the sample, recruitment to the study, sexual behaviour with male and female partners, attitudes to and use of condoms, other health behaviours, attitudes to and experience of HIV antibody testing and social integration (Fitzpatrick, Hart et al 1989; Fitzpatrick, Boulton et al 1989; Fitzpatrick, Dawson et al 1989; Fitzpatrick, McLean et al 1990; Hart et al 1990; Dawson et al 1991). In this project, face-to-face interviews in the period mid-1987 to early 1989 were followed-up with self-completed questionnaires nine to twelve months later. Asking questions about sexual behaviour in the last month and year means that, like some of the North American studies, we can interrogate our findings using the concept of relapse.

Of the 502 men who were interviewed in the first phase of our study, 369 (74 per cent) returned completed questionnaires at follow-up, and of these 356 (71 per cent) have provided sufficient data for relapse analysis. Like Stall et al we found that the data from the previous month – although very useful for the detail it provided of recent sexual activity – did not allow meaningful analysis in relation to the change from 'safe' sex (i.e. no partners, no penetrative anal sex, or anal penetration with condoms) to 'unsafe sex' (penetrative anal sex without condoms). This was because some men who were 'safe' at each of the two months in question may have had unsafe sex in the year prior to each of the months, and this would be lost in the analysis of such limited time periods. The comparison was therefore of the year prior to each of the two waves of the study.

Comparing the year prior to the first interview to the year prior to the follow-up questionnaire, the majority of the sample either remained safe (141; 40 per cent) or changed from unsafe to safe behaviours, including having no sexual partners (57; 16 per cent). Over a third of the sample were having unprotected penetrative anal sex during both time periods (132; 37 per cent), and were therefore – in these terms – consistently unsafe. Finally, the group which is the focus of the entire enterprise, are the 26 men (7 per cent) who changed from safe to unsafe behaviours during the two periods of comparison, i.e. those who 'relapsed'.

We have found few demographic, social, psychological or other behavioural characteristics of the 'relapsers' to distinguish them from the rest of our study population (Fitzpatrick et al 1991; Hart et al 1991). Across the population as a whole what we can say, like Stall et al, is that partner type is strongly associated with 'unsafe sex', with the majority of those men in regular relationships reporting unprotected penetrative sex with partners (58 per cent) as compared to those men with casual or non-regular partners (38 per cent), but no other clues as to any differences between the 'relapsers' and any of the other groups have so far been identified. These data do not 'disprove' the relapse thesis, and the relatively small sample size does not allow more rigorous analysis, but questions are raised as to the explanatory value of the concept.

Our findings may not be surprising. We have arbitarily taken two years of these men's lives and their sexual activity during this time and have then subjected this information to analysis in terms of maintenance and change over the period. These findings then are limited to the period of study; the descriptions one applies to the groups identified are not static and immutable, and men will change their sexual behaviour according to life-stage, relationships, strategies for maintaining health and acceptance of or response to given and known risk. The longer the period chosen, the higher the potential for movement within and between groups, and therefore the greater the likelihood of the appearance of 'relapse'. Relapse therefore becomes a function of time, and so it is not surprising that in a two year study the relapse rate was 4 per cent (O'Reilly et al 1990) whereas in the five year AIDS Behavioural Project it had reached 19 per cent (Stall et al 1990a).

What we are suggesting is that the concept of relapse may not prove empirically robust. This is not to say that in a technical sense the concept is an artefact of the data, but rather that in having been introduced *post hoc* it may appear to bear fruit in providing 'results' but could be describing – at best – variability in behaviour related to other, as yet unexplored, factors. That there are very real consequences of engaging in unsafe sexual behaviour cannot be doubted. Reports have appeared recently, to be discussed later, which suggest an increase in HIV seroconversions and in rectal gonorrhoea in gay men, and these must be explained. Before suggesting some means of understanding such developments, we will look at the concept of relapse from a different perspective.

'Relapse': conceptual critique

This is not simply a critique of the operationalization of a neutral concept. We are also concerned about the model of human sexuality employed in these studies, if unconsciously, and the ideological construction of the population to which relapse has been applied. Before questioning this episteme, it is worth describing the socio-linguistic basis of 'relapse'. We will end by suggesting what this means for a social construction of the gay community.

From a medical perspective it is understood that the expression of frank disease through signs and symptoms may be followed by a resolution of these, as a result of treatment or of natural history. However, some disease, particularly when this is systemic and/or chronic, is marked by remission (the period when the disease has subsided and florid symptoms are absent) which is a precursor to *relapse* (a return of disease after apparent resolution). Relapse is therefore a term used frequently in medicine, and particularly in psychiatric and psychological medicine in the treatment of alcohol and drug dependency, when periods of a patient's abstinence (when they are 'drug free') may be followed by a return to the use of the drug of addiction. Outside medicine, the term is more clearly pejorative, and refers to back-sliding, or slinking back to an unacceptable position.

Relapse, then, is concerned with a return to bad behaviour or state of being. The use of such disease derived and negative language for behaviour (which in itself is neutral until invested with meaning) reflects a medico-moral stance which many in the AIDS field have eschewed in favour of more positive language for discussions of sexuality and sexual activity. It should be noted that a disease model is not being applied to homosexuality per se in this instance, but to one dimension of sexual behaviour. However, that this has consequences for a construction of gay male sexuality will be made clear.

The question then arises as to the application of relapse to behaviour which is only associated with HIV when – obviously enough – HIV infection is present. In a medical setting relapse would have a clear meaning – the return of signs or symptoms – yet in the reports so far described it is the *threat* of infection which is present. Whilst the term 'prevention failures' used in one paper to describe men who had seroconverted may be mechanistic, it does refer to the presence of disease, rather than behaviour with risk attached to it. Sociologists are often concerned with context and meaning, and in this instance such a concern is very apposite; so much depends on the situation in which unprotected penetrative sex takes place, or a return to it, that invective against it can become sterile or even counter-productive at a health education level when the behaviour under investigation is without risk of infection.

However, a more important point relates to the nature of the activity to which men are said to relapse. The implication here is that the behaviour itself is to some extent addictive. Stall et al refer to:

> the difficulties associated with changing other health-compromising behaviours ... be they sobriety from alcohol or drugs, (or) adherence to a low-calorie or low-cholesterol diet. ... (1990b:1181).

Yet for men who share the same antibody status, and either do not have sex outside the relationship, or only have safe sex with partners outside the central relationship, unprotected anal intercourse is not something to which they have slid back, as one might to high cholesterol food, alcohol or drugs, but a chosen activity to enhance a particular relationship. That there are other men who engage in risk behaviours, regardless of the known possibility of HIV infection, has long been a feature of the epidemic, and clearly such behaviour needs to be explained. However, there is no evidence that addiction-influenced models have explanatory potential in this field for the majority of men affected.

We come now to the model of human sexual response upon which the assumption of relapse is predicated, and to an implicit social construction of gay men. In this formulation, relapse is a symptom of an internal sexual drive which must find expression through unprotected anal intercourse, unless some external control, which can then be internalized as self-discipline, can be applied. Sexual response is therefore equated to penetration, whether as inserter or receptor, and gay men in particular are over-determined as sexual beings (temporarily in remission) who will inevitably revert back to this behaviour. This construction of sexuality is more frequently associated with heterosexual men, and derives from nineteenth century interpretations of male sexual response, in terms of an overabundance of sexual energy which must be channelled into legitimate sexual activity (sexual relations in marriage, use of prostitutes) or into intellectual or physical effort (Weeks 1981; Mort 1989). What we see here, therefore, is an account of male sexuality – incorporating hetero and homosexual behaviour – which is dependent on penetration, urges or needs that are inherently difficult to control, and the inevitability of return (relapse) to given forms of sexual expression in the absence of thorough, consistent and long-term policing.

This conceptual admixture of disease presence, remission, back sliding, addiction and the inevitability of unrestricted penetration in male sexuality is a combination which, when applied to populations of gay men, further reinforces their otherness as outsiders marked by perversion (disease) and sexual appetites of heroic proportions. Such a social construction of gay men cannot be useful in any attempts to help them find ways to protect themselves from a life-threatening and sexually transmitted infection, and it is to this task that researchers in this field are expressly committed.

We will end by suggesting the reasons why the term 'relapse' has appeared at this time, the speed with which it was taken up, and the alternative methodological and conceptual means of approaching the whole issue of the prevention of HIV infection.

Conclusion

Longitudinal studies of gay male behaviour in response to HIV/AIDS have been underway in North America since the early 1980s, and in Europe and Australia since the mid-1980s. It is therefore interesting to note the widespread introduction of relapse into the analysis of longitudinal studies after 1989. For example, in that year a paper typical of current thought was published, which monitored sexual behaviour in a gay male population (McCusker et al 1989); relapse was not mentioned, although re-interpretation of the data would have produced a small number of men who could be located in this category.

Why should major research projects adopt this term, and why now? For many in the AIDS field, the magnitude of behavioural change reported in the gay population was such that it was considered one of the few 'success' stories associated with the epidemic. The message to health educators, health policy analysts and, finally and most importantly, politicians was that preventive behavioural changes were possible, and could determine the future course of the epidemics of HIV infection and AIDS. Whilst some have argued that there has been a paucity of funding in the AIDS field, in the USA and other developed countries large sums of money were directed to AIDS prevention programmes, usually from State but also from Federal funds. With the ongoing budget deficit crisis in the USA, however, a number of health and welfare programmes were threatened in the late 1980s. By 1989/90, and with the apparent success and extent of behavioural change in the gay population, politicians looking to save money could argue that health education for gay men had been money well spent, and resources should now be targeted to other 'at risk' populations. From a sociology of knowledge perspective, then, it is interesting to note that the concept of relapse should appear at the point at which health programmes for gay men were threatened with an uncertain future.

What we are *not* positing is that 'relapse' was introduced to save the jobs of health educators, community workers or researchers; we do not, like Bartley, subscribe to a narrow 'interest theory' to account for scientific developments and emphases (Bartley 1990). Indeed, we would go further and suggest that the only 'vested interest' here lies in highly committed researchers wishing to do whatever is in their power to prevent further infections. Stall et al in particular suggested the redirection of then existing and available monies, rather than arguing for continuation and renewal of resources. However, it *has* been necessary for workers in the field to put in the strongest terms the present and continuing risks that gay men face in relation to unprotected anal intercourse, and to press the case for the need for interventions which ensure not only that behavioural change takes place, but that this is maintained. The political context of conservative and cost-cutting Federal government in the late 1980s and early 1990s was not one which was conducive to such concerns, and so the arguments implicit in the relapse thesis regarding the future shape of the epidemic and the social, economic and health care costs attendant on further AIDS cases, may have once again placed these issues in the forefront of the minds of budget holders.

In this analysis, then, relapse not only has an empirical presence at a particular moment in the history of the epidemic, and a conceptual consistency in the minds of some who view gay male sexuality in disease terms anyway, but also a political topicality intimately connected to the perceived needs of the gay community. One cannot emphasize too strongly the consequences of redirecting health education efforts towards relapse. Current advertising for gay men in California challenges them to think up more 'excuses' for having 'unsafe sex' and to dial a health counsellor at a given telephone number to report these. This emphasis on individual culpability, excuses and mealy-mouthed avoidance of facing the reality of their own relapse could lead, if the concept proves unsustainable within the gay community, to a massive squandering of limited resources and a failure to reach those most obviously at risk.

Further factors also combine to give relapse appeal as an explanatory concept. That is, there have been a number of reports of recent infections and risk behaviour in young gay men in the United States (Kellog et al 1990; Hays et al 1991), of rectal gonorrhoea (usually taken as a proxy marker of unsafe sexual behaviour) in gay men in London (Singaratnam et al 1991) and Amsterdam (Van den Hoek et al 1991), and an increase in seroconversions in gay men in England (Waight and Miller 1991).

We do not usually know the HIV antibody status of men with rectal gonorrhoea, and so it is possible that this includes men of concordant antibody status. It has also been suggested of rectal gonorrhoea, that there are sexual means of transmission other than penile penetration of the anus (Tomlinson et al 1991; French et al 1991). However, assuming that unsafe sex between partners of unknown antibody status is occurring, could this and other evidence of risk behaviour and infections be explained by reference to relapse?

So far none of these papers demonstrate that the men becoming infected or at risk have returned to unsafe behaviours. It appears that some young men may have begun their sexual lives by having unsafe sex. In the studies of men seroconverting, the only distinguishing feature at present also appears to be younger age. It is so far not clear whether men who have repeated tests for anti-HIV – and thereby are included in seroconversion studies – are themselves monitoring continued risk-taking or single or multiple instances of 'relapse'. Some may have decided to be tested due to the availability of improved treatments, are found to be positive and thereby increase the incidence of seroconversion. Studies clearly need to determine why, at this stage of the epidemic, there are still seroconversions amongst gay men, and further robust efforts to prevent these.

We are arguing that the use of the term 'relapse' to explain recent developments in the epidemic is not indicated or supported as yet. Indeed such usage may actively work to the disadvantage of the gay community or sections of it. This would occur if, as we suggest here, the term fails to stand up to sustained methodological scrutiny and as an empirically flawed concept, researchers are later shown to have been 'crying wolf' in the absence of a second wave of infections amongst those members of the gay male population who have ostensibly relapsed. There is also the problem of the use of negatively laden terms which perpetuate a social construction of the gay male community in sexually reductionist language and conceptualization.

What we would suggest is that rather than rely on static, mechanistic and purely behavioural notions such as relapse or prevention failure, studies should actively address, using process-oriented analyses, the strategies employed by gay men in different contexts and with different partners either to avoid HIV infection or to live with the consequences of their risk-taking. It is also important to recognize that some people will be open to persuasion that they should protect themselves and their partners, others will not; this too must be accepted. Studies might start by asking men about their protective strategies through direct questions and by taking note of perceptions of partners, the affective, emotional and social context of sex, and by taking cognisance of local and situational features related to sex and HIV. This unfortunately does not have the immediacy and simplicity of the notion of relapse, but understanding the complexity of gay male sexual behaviour is not necessarily helped by the application of reductionist concepts.

It is worth concluding by re-affirming that many of the people actively involved in studies of gay men are highly committed to securing the best possible services for the gay community, and are passionate in their support for people living with HIV and AIDS. It was most moving when, at the Sixth International Conference on AIDS, one of the key researchers in this field made public his own diagnosis of HIV disease, and thereby brought to life the intimacy and coterminosity of public and private concerns in the epidemic. Our argument here is not with these researchers as such, but rather with the path down which concepts such as relapse might take us. It is our view that there is no inevitability about a 'return' to unsafe sex practices in the gay community, and that other means exist for ensuring that gay men continue to enjoy robust sexual and physical health.

References

Adib, M., Joseph, J. and Ostrow, D. (1990), 'Relapse in safer sexual practices among homosexual men: two year follow-up from the Chicago/MACS', Abstract FC724, *Proceedings of the VI International Conference on AIDS*, San Francisco.

Ames, L. and Beeker, C. (1990), 'Gay men in small cities: how risky are they?', Abstract FC719, *Proceedings of the VI International Conference on AIDS*, San Francisco.

Bartholow, B., Chon, D., Cole, V., Judson, F. and O'Reilly, K. (1990), 'From low-risk to high-risk: distinguishing among groups of gay men on the HIV relapse continuum', Abstract FC718, *Proceedings of the VI International Conference on AIDS*, San Francisco.

Bartley, M. (1990), 'Do we need a strong programme in medical sociology?', *Sociology of Health and Illness*, **12**:371–90.

Dawson, J., Fitzpatrick R., McLean J., Hart G. and Boulton, M. (1991), 'The HIV

test and sexual behaviour in a sample of homosexually active men', *Social Science and Medicine*, **32**:683–8,
Ekstrand, M.L., Stall, R.D., Coates, T.J. and McKusick, L. (1989), 'Risky sex relapse, the next challenge for AIDS prevention programmes: the AIDS Behavioural Research Project', Abstract TD08, *Proceedings of the V International Conference on AIDS*, Montreal, Canada.
Ekstrand, M.L. and Coates, T.J. (1990), 'Maintenance of safer sex behaviors and predictors of risky sex: the San Francisco Men's Health Study', *American Journal of Public Health*, **90**:973–7.
Evans, B.A., McLean, K.A., Dawson, S.G., Teece, S.A., Bond, R.A., MacRae, K.D. and Thorp, R.W. (1989), 'Trends in sexual behaviour and risk factors for HIV infection among homosexual men, 1984–7', *British Medical Journal*, **298**:215–18.
Fitzpatrick, R., Boulton, M. and Hart, G. (1989), 'Gay men's sexual behaviour in response to AIDS – insights and problems', in Aggleton, P., Hart, G. and Davies, P. (eds), *AIDS: Social Representations, Social Practices*, Lewes: Falmer.
Fitzpatrick, R., Boulton, M., Hart, G., Dawson, J. and McLean, J. (1989a), 'High risk sexual behaviour and condom use in a sample of homosexual and bisexual men', *Health Trends*, **21**:76–9.
Fitzpatrick, R., Dawson, J., McLean, J., Hart, G. and Boulton, M. (1989b), 'The life-styles and health behaviours of gay men', *Health Education Journal*, **48**:131–3.
Fitzpatrick, R., Hart, G., Boulton, M., McLean, J. and Dawson, J. (1989c), 'Heterosexual sexual behaviour in a sample of homosexually active men', *Genito Urinary Medicine*, **65**:259–62.
Fitzpatrick, R., McLean, J., Dawson, J., Boulton, M. and Hart, G. (1990), 'Factors influencing condom use in homosexually active men', *Genito Urinary Medicine*, **66**:346–50.
Fitzpatrick, R., Dawson, J., Boulton, M., McLean, J. and Hart, G. (1991), 'Social psychological factors that may predict high risk sexual behaviour in gay men', *Health Education Journal*, **50**:63–6.
French, P.D., Mercey, D.E., Tomlinson, D.R. and Harris, J.R.W. (1991), 'Preventing the spread of HIV infection', *British Medical Journal*, **302**:962.
Hart, G. (1989), 'AIDS, homosexual men and behavioural change', in Martin, C.J. and McQueen, D.V. (eds), *Readings for a New Public Health*, Edinburgh University Press.
Hart, G., Fitzpatrick, R., McLean, J., Dawson, J. and Boulton, M. (1990), 'Gay men, social support and HIV disease: a study of social integration in the gay community', *AIDS Care*, **2**:163–70.
Hart, G., McLean, J., Boulton, M., Dawson, J. and Fitzpatrick, R. (1991), 'Maintenance and change in safer sex behaviours in a cohort of gay men in England', poster presented at the VII International Conference on AIDS, Florence.
Hays, R.B., Kegeles, S. and Coates, T. (1991), 'Understanding the high rates of HIV risk-taking among young gay and bisexual men: the young men's survey', Abstract MC101, *Proceedings of the VII International Conference on AIDS*, Florence.
Holland, J., Ramazanoglu, C., Scott, S., Sharpe, S. and Thomson, R. (1991), 'Between embarrassment and trust: young women and the diversity of condom

use', in Aggleton, P., Hart, G. and Davies, P. (eds), *AIDS: Responses, Interventions and Care*, Lewes: Falmer.
Johnson, A.M. and Gill, O.N. (1989), 'Evidence for recent changes in sexual behaviour in homosexual men in England and Wales', *Philosophical Transaction of the Royal Society of London*, **B325**:153–61.
Kellog, T.A., Marelich, W.D., Wilson, M.J., Lemp, G.F., Bolan, G. and Rutherford, G.W. (1990), 'HIV prevalence among homosexual and bisexual men in the San Francisco Bay Area: evidence of infection among young men', Abstract FC548, *Proceedings of the VI International Conference on AIDS*, San Francisco, California.
Kelly, J.A., St Lawrence, J.S. and Brasfield, T.L. (1991), 'Predictors of vulnerability to AIDS risk behaviour relapse', *Journal of Consulting and Clinical Psychology*, **59**:163–6.
Loveday, C., Pomeroy, L., Weller, I.V.D., Quirk, J., Hawkins, A., Smith, A., Williams, P., Tedder, R.S. and Adler, M.W. (1989), 'Human immunodeficiency viruses in patients attending a sexually transmitted disease clinic in London, 1982–7', *British Medical Journal*, **298**:419–21.
McCusker, J., Stoddard, A.M., Zapka, J.G., Zorn, M. and Mayer, K.H. (1989), 'Predictors of AIDS-preventive behaviour among homosexually active men: a longitudinal study', *AIDS*, **3**:443–8.
Mort, F. (1989), *Dangerous Sexualities: Medico-Moral Politics in England Since 1830*, London: Routledge and Kegan Paul.
O'Reilly, K.R., Higgins, D.L., Galavotti, C. and Sheridan, J. (1990), 'Relapse from safer sex among homosexual men: evidence from four cohorts in the AIDS community demonstration projects', Abstract FC717, *Proceedings of the VI International Conference on AIDS*, San Francisco.
Paton, C. (1989), 'Resistance and the erotic', in Aggleton, P., Hart, G. and Davies, P. (eds), *AIDS: Social Representations, Social Practices*, Lewes: Falmer.
Richardson, D. (1990), 'AIDS education and women: sexual and reproductive issues', in Aggleton, P., Davies, P. and Hart, G. (eds), *AIDS: Individual, Cultural and Policy Dimensions*, Lewes: Falmer.
Rosenstock, K. (1974a), 'Historical origins of the Health Belief Model', *Health Education Monthly*, **2**:328–35.
Rosenstock, K. (1974b), 'The Health Belief Model and preventive health behaviour', *Health Education Monthly*, **2**:354–65.
St Lawrence, J.S., Brasfield, T.L. and Kelly, J.A. (1990), 'Factors which predict relapse to unsafe sex by gay men', Abstract FC725, *Proceedings of the VI International Conference on AIDS*, San Francisco.
Singaratnam, A.E., Boag, F., Barton, S.E., Hawkins, D.A. and Lawrence, D.A. (1991), 'Preventing the spread of HIV infection', *British Medical Journal*, **302**:469.
Stall, R. Ekstrand, M., Pollack, L. and Coates, T.J. (1990a), 'Relapse from safer sex: the AIDS Behavioural Research Project', Abstract ThC108, *Proceedings of the VI International Conference on AIDS*, San Francisco.
Stall, R., Ekstrand, M., Pollack, L., McKusick, L. and Coates, T.J. (1990b), 'Relapse from safer sex: the next challenge for AIDS prevention efforts', *Journal of Acquired Immune Deficiency Sydromes*, **3**:1181–7.

Tomlinson, D.R., French, P.D., Harris, J.R.W. and Mercey, D.E. (1991), 'Does rectal gonorrhoea reflect unsafe sex?', *Lancet*, **337**:501–2.

Van den Hoek, J.A.R., Griensven, G.J.P. and Coutinho, R.A. (1990), 'Increase in unsafe homosexual behaviour', *Lancet*, **336**:179–80.

Waight, P.A. and Miller, E. (1991), 'Incidence of HIV infection among homosexual men', *British Medical Journal*, **303**:311.

Wallston, K. and Wallston, B. (1981), 'Health locus of control scales', in Lefcourt, H. (ed.), *Research with the Locus of Control Construct*, New York: Academic Press.

Weeks, J. (1981), *Sex, Politics and Society*, London: Longman.

Willoughby, B.C., Schechter, M.T., Craib, K.J.P., McLeod, W.A., Douglas, B., Fay, S., Nitz, R. and O'Shaughnessy, M. (1990), 'Characteristics of recent seroconverters in a cohort of homosexual men: who are the prevention failures?', Abstract FC45, *Proceedings of the VI International Conference on AIDS*, San Francisco.

Wilton, T. and Aggleton, P. (1991), 'Condoms, coercion and control: Heterosexuality and the limits to HIV/AIDS education', in Aggleton, P., Hart, G. and Davies, P. (eds), *AIDS: Responses, Interventions and Care*, Lewes: Falmer.

10 Quo vadis the special hospitals?

Joel Richman and Tom Mason

Introduction

Special Hospitals are those maintained under Section 4 of the NHS Act, 1977 (formerly section 97 of the Mental Health Act, 1959), whereby the Secretary for Social Service oversees them. Special Hospitals are designated for those detained under the 1983 Mental Health Act, whom 'the Secretary' considers requiring 'treatment under conditions of special security (category A) on account of their dangerousness, violent or criminal propensities'.

Dangerousness is historically rooted. In the early nineteenth century those crimes attributed to insanity became crimes against nature, not against the laws of society. As Foucault (1978) noted, criminal psychiatry first proclaimed itself a 'pathology of the monstrous'. These monsters, later regarded as a sub-human species in Darwinian mode, were dangerous because their insanity and their criminality fused to render their crimes unintelligible in terms of reason, profit or emotion. These 'evolutionary misfits', not susceptible to cure and reform, could not be lodged in mental institutions and prisons (the latter undergoing reform) lest they contaminated the 'normal' deviants. Special Hospitals became the ultimate repository for the 'uncontrollable'. Today some forensic psychiatrists accrue infamous patients, claiming a degree of treatability, but gaining fame based on monstrous notoriety.

Today, dangerousness in all its aspects culturally loads the Special Hospitals. Nursing staff new to a ward will be briefed by colleagues on who are 'hostage takers'; student nurses are constantly reminded to be within visibility of staff; patients' possessions undergo random searches (poisoners can cultivate dangerous moulds from food); psychiatric wisdom has it that the 'overcontrolled' (quiet) psychopath is the most dangerous, erupting when least expected.

The official definition reveals nothing about the social processes within these hospitals, or their relationship with the wider penal system and other NHS psychiatric facilities. There are now three Special Hospitals in England: Broadmoor built with convict labour, 1863: Rampton 1912, Moss Side 1919 and Park Lane (adjacent to Moss Side) 1974. Park Lane was founded with seventy male patients transferred from Broadmoor as an advance unit with the official opening in 1984. Park Lane and Moss Side amalgamated in 1990 to form Ashworth Hospital – the name being the result of a hospital competition. In 1989 there were 1700 patients in the three Special Hospitals.

Reflecting the structural ambiguities and philosophy of the Special Hospitals, there has been the inter-site movement of patients – Moss Side patients were transferred to Rampton in 1920. Some Rampton patients who gave evidence to the Boynton enquiry, 1979, investigating allegations of staff brutality were also moved out. Changes in national policy, penal and psychiatric ideologies, together with public moral panics, have buffeted the 'Specials'. The biographical shunting of patients like Frank Mitchell is indicative. On July 18, 1955, while in prison, he was certified 'mentally defective', on August 29, 1955, 'being feeble minded', was sent to Rampton. On January 18, 1957, he escaped and committed severe offences, including attempted murder and was sentenced to nine years imprisonment. He was then certified insane and sent to Broadmoor on April 18, 1957, from which he escaped in July 1958, and committed further offences and was sentenced to life imprisonment. The Medical Superintendent of Broadmoor and prison doctors found him fit to plead, arguing that his treatment in Broadmoor had helped him 'mature'.

More significant are the 15 per cent of patients in the Special Hospital system who have been waiting two years to move to NHS facilities. The interplay between political control and clinical care has never been reconciled; for example, in this case study the person who severely attacked a member of the judiciary was sent to the most secure unit in the Special Hospital. He had to be 'specialled' by at least three nurses, detracting from the care of other highly dependent patients. The Queen's unwelcome visitor was also temporarily lodged in a Special Hospital. Evading Buckingham Palaces's security, he conversed with the Queen in her bedroom. Patients sectioned under 39–41 have no EDR (estimated date of release), as they would have for comparable offences in prison. The patients' time vacuum influences their behaviour, often confusing it with illness symptomatology. Institutionalization often occurs when meaningful, temporal markers 'disappear' (Richman 1989). In the 1980s, some patients had their medical progress reviewed annually. A patient can resort to the strategy of 'self degradation' as one of the options available for identity maintenance. Returning to the ward norms of cleanliness provides a visible indicator of 'progress'.

To outsiders, Special Hospitals are secret societies. There are few comprehensive studies of them. The public have been fed on a diet of bad news ... prefaced by 'containing the most dangerous men in England'. Recently this gender bias was broken down: 25 per cent of patients are women – the former Park Lane was the only all male 'Special' – and 1990 saw the first conference focusing on women in Special Hospitals. Their double transgression – against the law and against their own 'natures' – was highlighted as the rationale for many being referred to the 'Specials'.

Sections of the national press have storylines prepared when some patients are released, or about to go before the Mental Health Review Tribunal. Nursing staff report that they have been approached by the press for information regarding the more notorious and politically sensitive patients, whose release would endanger 'national morality'. For over twenty-five years the 'Moors Murderers', for example, have been continuously portrayed as totemic of 'absolute evil'. The press had expressed delight when it was alleged that one notorious patient had lost money on the BP share flotation. The notion of dangerousness can be traded upon by nursing staff as a justification for restricting care or innovation. The multidisciplinary approach to treatment can lead to professional rivalries, as each specialism jostles for the hierarchical validity of its 'approach'. Nurses and others, to prevent their role attrition, will block such moves – not on the grounds of efficacy but by evoking the new risks contained in new clinical practices. The high wall surrounding the hospital is the conspicuous reminder to all of 'danger-containment'.

Patients undergo multiple 'taintedness', firstly, by the Mental Health Act sections (not all come under them); by the Special Hospital tag itself; and by their ward location within the 'Special'. The Rampton report (1980) commented on the everyday life of the low status, Drake block ward. These patients rarely left the ward and 'cleaning occupies the whole morning period and is performed to a standard usually seen in operating theatres' (p 59). But patients in the more prestigious villa wards had more possessions, domesticated their rooms and had more rapport with staff. Some patients neutralize their pollution by scheming (often aided by their psychiatrist), to be transferred to prison for release. One former poisoner argued successfully at his tribunal that the Special Hospital could not treat him.

Regional Secure Units (RSUs) are reluctant to accept certain types of 'Special' patients because of their own internal philosophies and past experiences. General psychiatric hospitals are even more reluctant to accept 'Special' patients and a crude indicator of the insularity of the 'Specials' frequently emerges in nursing talk with some still referring to NHS hospitals as 'county hospitals'.

This chapter has a number of broad aims:

(i) To adumbrate the psychiatric forensic tradition, thereby explicating the major research thrust underpinning care and offering types of explanations.

(ii) To elaborate the existing 'culture of care' and pose what is now known as the 'psychopath issue'.

(iii) To explore recent nurse-education changes and list outline effects (the 'Specials'' own RMN nurse training programme is under threat of closure forcing the schools to develop wider nursing links and diversify their ethos).

(iv) To examine forces resistant to organizational change and express the newly emerging administrative structure, and to signpost what may be considered as the 'final' scenarios for the 'Specials'.

The analysis pertains to one Special Hospital, with comparative reference to the others, easily distinguished in the account.

Forensic tradition

This can be characterized as atomistic, thematic, policy-driven (without explicating the tacit assumptions and values involved in being the 'servants of power'), genetic, biological and personality hinged – but largely devoid of ethnographic texturing, especially from inside the Special Hospital. It is paradoxical that low-status student nurses, as part of their qualification, produce projects involving 'culture slices' of ward behaviour etc, matrixing psychiatric practices. For example, much endeavour is invested in explaining, diagnosing and predicting 'dangerousness' (and anti-social behaviour). The latter being continually refined in the mill of semantics and legal fiat. Recently, Hamilton and Bullard (1990) (dismissing without comment the Home Office and DHSS (1975) Committee on Mental Abnormal Offenders definition: 'a propensity to cause serious physical injury or lasting psychological harm') disagree with Scott (1977) that 'dangerousness is an unpredictable and untreatable tendency to inflict or risk serious, irreversible injury or destruction', and shuffle their version of dangerousness through a catalogue of rape, paedophilia, exhibitionism, arson, schizophrenia, manic-depressive psychosis and so on, evolving with a recipe knowledge conclusion. Any action can be designated as dangerous – even ultra conformism. Gunn (1985) is open, 'We learn yet again that our ability to forecast which people will do harm is very poor'.

As illustrative of the psychiatric methodology examining 'violent incidents' within Special Hospitals, Coldwell and Naismith (1989) log and ecologically map 116 violent incidents over a twelve month period in two special care wards in Park Lane. The figures are the highest published. After listing demography, index offence, diagnosis, medications, authority for detention and ward lay-out, they reference violent incidents (after constructing their three-fold types). They show that only 6 per cent of incidents were recorded between 2130 and 0730 hours (patients are locked in separate rooms during this period): 60 per cent between 1400 and 2130 hours, in which period the authors suggest more patients are present in the communal ward area. However, other explanations are avoided, the authors do not explain why they compressed two separate wards into one data base. Staff on ward B regard themselves as superior to A and did not 'hit the bell' (alarm system) as frequently.

The official documentation of what constitutes an 'incident' is socially constructed. When staff record that their watches were broken or shirts torn, for example, the official reader is guided to the conclusion that this was a 'really' violent incident. The authors do not comment on some data; 17 per cent of incidents occurred within seclusion rooms. These special care wards, isolated within the hospital, receive patients who had attacked staff elsewhere in the hospital (amongst other things) and were known in staff and patient argot as the 'punishment block', the 'sink', or the 'Bermuda Triangle'.

Patients were also put there for their own safety, from patients on other wards; or because they were self mutilators (which is more common in female patients); or housed there temporarily if staff from other wards did not want them to participate in significant ward events. Some patients on these wards were on medication dosages not within BNF (British National Formulary) limits. Incidents rise on Sundays, but no comment is made of the use of tobacco on these wards – despite staff rationing, tobacco is running out before 'canteen day' on Mondays. Tobacco has multipurposes. Smoking creates a personal timetable over which patients exert some control. Being hunched over a cigarette often signals the requests to be 'left alone'. Addiction is very high and those with no cigarettes can intimidate others known to have them, precipitating incidents.

Three nursing groups rotate on each ward – each with different styles. The authors do not attempt the obvious – to see if incidents cluster more with a given staff grouping. At the time of the study, patients were not allowed to filter to their rooms during the day. They were instead transformed into very public people, especially those who are easily overwhelmed by heavy interpersonal contact, some 'turn off in their heads' and others attempt to get 'time out'. Finally, what the author omits to tell the reader is that as the psychiatrist in charge of one of the wards under study he is built into the context.

The psychopath issue

The issue centres on two concerns: *(a) does such a 'thing' exist and (b) if it does/does not, can it be treated?*

Despite extreme vagueness, the category persists, for example the Butler Report (1975):

> *multiplicity* of opinions as to the aetiology, symptoms and treatment of 'psychopathy', which is only to be understood by reference to the *particular* sense in which the term is being employed by *the* psychiatrist in question,

this resulting in a *free enterprise* diagnosis. *Offenders Suffering from Psychopathic Disorder* (DHSS/Home Office Consultation Document, 1986),

> is not a description of a single clinical disorder but a *convenient* label to describe a severe personality which may show itself in a variety of attitudinal, emotional and inter-personal behaviour problems (p 4)

resulting in an *Azande evidence* diagnosis. Evans-Pritchard's (1937) magnificent study of Azande magic, a closed system of belief, elaborated how any 'unusual' phenomenon can be transformed into proof of witchcraft. Similarly any aspect of human behaviour has the potential to be read as evidence of psychopathology, once the residual assumption has been established that psychopaths exist.

The debate as to whether there is any clinical value in treating psychopaths in 'Specials' is long standing. The Royal Medico-Psychological Association memorandum (1968) argued that they should go to prison 'unless manifestly complicated by mental illness or severe subnormality'.

To move psychopaths en masse to prison would have serious consequences to the existing prison reform schemes, unless the courts award more 'community orders' to other offenders. The term sociopath is gaining currency, partly following the US tradition, as one who mouths fine words but commits horrendous offences. As no common genetic cause has been discovered for their behaviour, explanations of psychopaths are now drifting towards 'defective' socialization.

Within the Special Hospital, a special care unit was set up specifically for the psychopathic patient. Officially it is just a ward, like any other ward, because of their former neglect within the clinical regimes. This raises the fundamental question *how can you treat the 'untreatable', or that which does not exist?* This leads to a number of consequences, firstly, an increased amount of 'clinical attention' was directed towards them, reducing the potential elsewhere. Secondly, the nursing staff could not adopt their usual orientations. Explanations had to be offered for all their actions, that is, 'nursing with moral justifications'. This soon exposed the weakness that the nursing role was intellectually unsustainable. The unit soon accrued the highest ranking and privileges in the patient hierarchy and became the benchmark for other wards to achieve. This made more problematic the 'frontier of control' within the 'Special'. Patients were accorded the freedom/

rights to 'act-out' personal and group troubles, provided there was no physical violence. What was a seclusion offence elsewhere was not necessarily so on the unit. Other patients and staff regarded this as unfair and the future direction of the unit is still unclear.

Nursing and culture

A corollary of the 'medicalization' of criminology has been the shadowy emergence of the nursing and prison officer roles fused from both those preceding professions into the Special Hospital nurse. The encapsulation of these two roles at ward level has engaged the two polarized positions in an ongoing mutual 'assault' of each other with a system of ridicule, humour and social distancing to reaffirm each particular value structure. Whilst some nurses are content to occupy each of the roles at the extreme poles, some attempts by others have been made to establish a 'middle ground', narrowing the parameters between 'nurse' and 'prison officer' in the belief that a central area is both extant and attainable. However, in working practice, the parameters of the roles appear to be narrowed, not to a single entity, but to reduced parameters whereby a degree of oscillation between the two occurs.

The uniqueness of the Special Hospitals in their development, their role and their clientele, and the uniqueness of the conjoint 'nursing' and 'prison officer' roles, has paved the way for the development of an atypical culture. This culture has several components, firstly the patients, the majority of whom arrive at the Special Hospital 'criminalized' by the system of *crime = police = court = punishment*. Whether or not they go to prison first (which a third do) or whether they are admitted direct to the Special Hospital, all undergo a process of 'dangerousness' assessment geared towards 'criminalization'. Some patients pay lip service to psychiatry engaging in 'therapeutic' activities as a pay off for the 'soft time' option that a Special Hospital is perceived by some to be, whilst others perpetuate their defence of admission to a psychiatric hospital as a 'con' However, there are still others so chronically ill or handicapped that they have spiralled their way down through various medical and judicial systems and merely have no other suitable institution available. It is also fair to state that some do react to the treatment model and are ultimately discharged to successful futures.

A second component is a treatment structure whose models, however, produce little response from the clients on which they are applied, be it from doctors, psychologists, social workers or nurses, and whose prognostic and predictive components are generally arbitrary. The medical model fails to match up to the twentieth century assessment and treatment of 'dangerousness' with resultant 'professional' demoralization and subsequent 'safe' maintenance of patients in the Special Hospital system. Statistically only one third of released patients will re-offend. The difficulty is identifying which third and why!

Thirdly, the medicalization process is supported by a battery of nursing values, propounded with their usual semi-religious fervour and perpetuated by the traditional nursing contractualism implicit in patient sick-role adoption, care plan negotiation and treatment efficacy. However, this framework clashes with the cultural components outlined above as the inability to achieve mutual goals becomes apparent and the patient having undergone the 'treatment' process is still considered dangerous and thus detained further with ever-increasing frustration on all sides. The neutralized therapeutic efficacy results in a 'provision of services' approach to fill the time vacuum that remains. This extremely expensive containment is being closely scrutinized by both public and politicians and there is increasing pressure to evolve efficient clinical indicators in the construction of the forensic knowledge base.

Finally, the cultural mosaic consists of the security aspect of the Special Hospital's function which remains the fundamental emphasis, and the reaction to breaches of security are a good indication of the prioritization of values. Nursing staff receive pressure to provide a nursing service which is based on care and compassion, and to form nurse–patient relationships based on trust and negotiation, and are covertly expected to be 'risk-takers' in their attempts to provide a therapeutic service. However, managerial response to staff, who are deemed responsible for infringement of security, receive a disproportionate punitive sanction than breaches in, say, provision of 'therapy'. This ensures staff emphasize the security consideration in their daily applications. It is interesting to note that the new Special Hospitals Service Authority, which has been operational for approximately three years, is considering a support framework for staff who take 'professional' risks in endeavouring to establish a treatment basis for their activities, which should help lessen the emphasized focus of attention on 'security'.

Developing from these various pressures, the Special Hospital nurse stands in a unique sub-culture and attempts to fulfil the medical promise, the nursing contract and the societal 'policing', in a type of order-conflict dimension. According to context, the nurse's role will oscillate between heavy control in quelling violence and in caring. The nursing 'contract' with the patient is officially based on administering the medical model and yet being the patient's advocate. When the promise of the medical model is unfulfilled the contract becomes 'distorted'. Nurses are then obliged to probe deeply into the patient's biography and everyday ward behaviour for signs and symptoms (not just of dangerousness), in order to perpetuate their detention on the hazy rationale that it is ultimately in the patient's best interest not to be released, due to the likelihood of re-offending.

Some wards more than others have eagerly embraced the nursing process/models of care – to refuse would be a direct challenge to the new managerialism installed over the last few years. The choice of model was often ad hoc; certainly installed without prior testing and modification. Paradoxically, on a high dependency (critical care) ward which more seriously deployed its chosen model, unintended consequences emerged. Patients who had satisfied the criteria of the ward model had their expectations highly raised about their future status passage. They assumed they now had an automatic right to move either to a medium dependency ward with a freer and less stigmatizing milieu or be transferred out of the hospital. However, logistics dictate that patients move when 'beds' become available, irrespective of clinical or social considerations. Psychiatrists or PCTMs (patient care team meeting) found themselves under 'new pressures' from the new ward validations. Patients had legitimate discontents when their 'respectable validation' did not produce the progress expected. Nursing staff's marginality increased; on the one hand they were now seen to be doing authentic nursing, but on the other were left to handle increasing patient dissatisfaction! The end stage of a nursing model did not equate with cure or return to normality (as it would if deployed in a general psychiatric hospital) and hence automatic release. An accumulation of thus dissatisfied patients, willingly offering cooperation to discover there was no immediate 'reward', increases frustration, making 'risk management' more problematic.

Staff have long used 'distancing techniques' to inhibit patients from gaining knowledge about their private lives. Patients become empowered when they have infiltrated nurses' extra-professional roles, especially family details. Nursing groups maintain a tight frontier of control. Staff's self revealment in therapy relationships would also expose the other side of masculinity – a softer and less rational one. To run wards on therapeutic lines needs a creative intellectual environment, involving a new style of nursing and one being much more people-centred; and of course, this is much harder work, with greater risks.

Effects of reorganization of psychiatric nursing

Within the continuous medico-legal non-alignment, power bases are expanded and retracted, as both sides ascertain strategic features of the other, reconnoitre territories, establish areas of incorporation and identify weaknesses for exploitation. With ground lost from the late 1970s and early 1980s media and legal onslaught in a series of television programmes and newspaper articles which led to litigation and conviction of some nursing staff, the Special Hospitals reeled under threat of abolition of nurse training and the possibility of closure. Attempting to regain lost ground or, probably more accurately, fighting for their very survival, the Special Hospitals responded by 'sterilizing' criticisms of secrecy with an almost 'open to everyone' policy.

The Specials clutched at the NHS 'straws'. The 1982 Nurse Training Syllabus, which despite attempts at experiential approaches was fundamentally based on behaviouralistic principles (a left-over from the 1970s), was implemented with little re-training or education of ward staff. The eclectic basis of the 1982 syllabus which dictated that a nurse is all things to all people, a jack-of-all-trades and a generalist, did not fit well with the 'Special' ideology. Student psychiatric nurses empowered with notions of 'questioning', 'enquiry' and 'criticism' were placed in a culture of traditional order and discipline. The result of this 'educated' student nurse placement in Special Hospitals has led to a conflict of values. This framework results in pressure on groups of people, mainly student nurses, to employ varying tactics to confront the established culture. However, the personal cost for some individuals who attempted to modify some 'bad' nursing practices has been high.

Project 2000 also had its impact on the Special Hospitals with the proposed move of nursing education to higher educational establishments. Each of the Specials is now aligned to such an institute in each of their localities. The underlying principles of Project 2000 are attempts to 'professionalize' nursing along the lines of the medical profession and, although it is well understood that this elitist approach will lead to a shortage of qualified nurses, this is combatted by the introduction of the Health Care Worker, trained to a specified standard, which will create the very two-tier, two-portals of entry system that the regrading and conversion course programme has been employed to eradicate.

For the Special Hospital schools of nursing, Project 2000 has re-defined their role towards planning for In-service Training facilities, staff development projects and post-basic ENB curricula, and encourages courses to be self-financing, by attracting external candidates which also serves the 'consumerism' approach to health care. However, perhaps the more dramatic corollary of attracting external candidates is the influx of alternative perspectives into the Special Hospital system from workers in the wide variety of nursing fields. Yet paradoxically, this failed to happen at the inception of Park Lane in 1974 with the recruitment of nurses from the NHS, possibly due to the overall management structure having been apprenticed at Broadmoor.

The Special Hospitals enmeshment in a fight for survival led to their taking on board any piece of armoury which would deflect criticism, however ineffective they were at creating change at ward levels (Mason and Chandley 1990), which was hastily adopted because of a crisis of confidence. It included two attempts at establishing the Nursing Process, both resulting in 'paper exercises', adoption of nursing models of care clearly not designed for the forensic setting, and an 'entertainments' approach to forensic nursing off-setting the impotence of the therapeutic model. (When one recent primary nurse was asked about the content of their care plan he retorted 'care plans are like the menus, full of stuff they never get'.) The Nursing Process was used at a practical and symbolic level; practical in that it gave the organizational hierarchy the structure to say that developments were being made, and symbolic in that 'professional' alignment to the wider nursing field could be 'seen'. Nursing staff considered 'good' at care plan construction became 'script writers' for the less able and were referred to in cases of 'difficult' patients in order to construct a care plan with the least 'loopholes'. The conceptual and theoretical understanding and appreciation of the ideas adopted were less important than their being 'seen' to have been so.

The stimulation to embrace the 'professional' values that underpin the 'ideas' of the Nursing Process, models of care, etc, did not appear to come into conflict with the trend in the 1980s towards managerialist values and the concomitant nursing audits, performance indicators and budgetary controls. Two possible reasons for this are that firstly, conflict does not exist because the former values are not dominant in the system, and secondly, because of the expected uncritical acceptance of change so apparent throughout the nursing discipline.

Although the consequences of psychiatric nurse education reforms are still resonant, some pointers emerge. Firstly, wards which had long basked in a routinized, task-orientated and theoretical style of nursing now had to declare openly, via some debate, but mainly documentation-wise, what were or should be their clinical objectives. As the key script writers emerged they presented their newly found philosophies to curriculum development groups of the new school of nursing, called the Elizabeth Campbell. Ward objectives and audits were necessary before some wards were declared acceptable as learning/training wards for the new student nurses. Many students soon found aspects to be legal-fictions, at best 'over-idealistic', producing learning confusions. The declaration of the first list of training wards radically disturbed the 'insiders' taken-for-granted ward rankings, based on other criteria. Self-ascribed, high-status wards not included in the education programme now competitively sought learner status to consolidate their culture and ranking. Thus a wave of artificially created clinical and nursing care semiotics emerged, further obfuscating and camouflaging ongoing reality.

Those student nurses who expressed a desire to nurse at the Special Hospital, after qualifying, did so mainly on the grounds of travelling convenience and for other instrumental reasons, such as the Special Hospital financial lead.

Some female learners expressed their own difficulties. The study of staff attitudes conducted prior to the ingress of the new students revealed ambiguities towards female learners. Some welcomed them, for they were expected to 'normalize' and 'soften' the tough masculine milieu; while other staff feared that they might 'lose control' and 'fraternize' unduly with male patients, thereby undermining the solidarity of the nursing group. It is interesting to note that no such fears were expressed about male learners, yet some nurses are known to be gay. Some female learners disliked how certain nursing groups would regard them as 'theirs'. In some instances the overprotective attitude merged with sexual harassment. Learners, generally, are powerless to change ward milieus (not just in Specials), especially in moderating frequent swearing. The female learner would be the only woman in an otherwise male nursing group.

The continuing attempts to 'prise' open the Special Hospitals through nursing education needs to be geared towards social understanding, as support is gained for accepting that individualistic paradigms are less appropriate for forensic psychiatry based on the belief that individuals are not isolational units but part of a social matrix.

Resistance to change

Many factors congeal to resist change. For many years the Special Hospitals, as closed institutions (or as some students call them, prison hospitals), have been beyond the pale of the wider developments in the NHS and education. They have originated their own traditions. It is noteworthy that when they have recently diffused knowledge 'elsewhere', dangerousness has been to the fore. The 'Specials' have pioneered control and restraint courses (C&R), based on selected joint holds less likely to damage patients. The English National Board is about to recognize the C and R course. Broadmoor has offered it to British Airways, for dealing with 'dangerous' passengers. (In fairness, the majority of nurses did not want the C&R courses; the latter also involves nurses operating in 'threes'; ward nurses traditionally handle incidents as a larger unified group.)

The 'Specials' fit Kerr's 'isolated mass' thesis, that is the close proximity and links between work site and community (as with traditional mining villages, which used to sustain long patterns of labour/management conflict)(Kerr and Siegel 1954). At Rampton, some nursing staff's connections extend back through generations. In 1980 the hospital was more a cottage industry, with staff working long shifts to suit themselves. The 1989 Review noticed progress, helped by a much reduced patient population, which decreased from 1070 patients in 1979 to 550 in 1989. The Review did comment, however, that 'what constitutes 'appropriate treatment' for patients needs to be clearly stated' (p 35). The Review was pleased that nurses were now more likely to talk about therapy than custodial issues.

Special Hospitals have always been top-down organizations, with real power lying outside with the 'Department' and 'Minister'. Policy direction is historically erratic and usually fault-finding, with punishment-centred rules for staff and patients alike. Thus the organizational climate was inimical for risk taking, an essential pre-requisite for innovation. So heavily engrained are legal procedures, that adjacent wards would correspond by 'letter'/memo, to establish 'the record'. To illustrate the organizational encumbrances, the Chief Nursing Officer of a Special Hospital could only suspend a nurse for severe unprofessional conduct, with the 'papers' ground slowly through the Department's machinery; his NHS counterpart, for the same misdemeanour, could instantly dismiss. Again, an enlightened Charge Nurse used his discretion and removed the ground parole card from a patient whom he accurately predicted was about to have a bad psychiatric episode. If the patient left the ward there was the strong probability that he would have broken the hospital rules and then 'knocked back' for progression. On recovery the patient was encouraged to complain about the Charge Nurse's 'dictatorial' stance and was reported to the Department. It took the Charge Nurse two years to clear his name. In summary, the organization of the 'Specials' could be described as patrimonial bureaucracy. Despite the thick layering of rules, the Chief Nursing Officer still dealt with a tremendous amount of minutiae fed up to him. This structure was incapable of responding to rapid change, translating it into coherent policies and successfully executing them with general consensus. This is now changing.

There was a large gulf between the wards and the rest of the administration. Psychiatrists and Nurse Managers spent comparatively little time on the wards. It was a common complaint of the nurses that doctors were never on hand to deal with violent incidents. The wards were enclosed time and space capsules for both patients and staff alike. Like other hospitals, the 'Specials' have spawned many professional groupings (excluding the nurse/medical line of control). Excluding admissions and critical care wards patients spent much of the day off the wards for occupational and other skills training, inhibiting the skills development of nurses (if they so wished). Nurses have increasingly found themselves with only their custodial role to fall back on. The repetitive rites of door locking/unlocking, serving meals and medication and 'banging up' at night, also condition nurses to acting in the *immediacy*. Boundary and knowledge disputes between the hosts of professions also detracted from projecting long term goals. The Broadmoor Review (1988) similarly commented on this.

Although nurses have never worn uniform, as in the other 'Specials', patients still refer to them as 'screws'. The Prison Officers Association, numerically small, is still influential in custodial settings. A few years ago the RCN (Royal College of Nursing), was not allowed within the walls to recruit staff. Management efforts are geared towards 'deinstitutionalization' of the POA to establish a framework for change.

Emerging organizational structure

Reorganization of the Special Hospitals in the 1980s was undertaken in a cauldron of power relations which has left the identification of the seat of power to be crystallized. Traditionally, the Special Hospitals had two distinct nursing organizational structures; the wards and the 'management'. Within each, a hierarchical/status framework existed which were interdependent from each other. Although staff were promoted from the wards to the 'management' as a continuum, they left the top of one hierarchy, in which the sphere of power could directly influence the ward, to move to the bottom of the 'management' hierarchy in which they had little influence in either the wards or management.

Accountability for the daily operations of the wards remained with the nurse in charge, who had almost total authority in relation to what happened to the patient – who saw them, when and for how long – and although the Responsible Medical Officer assumed overall responsibility, they had little say in the wards' structures, could be overruled by the Home Office in the case of discharge of 'dangerous' patients and had to work within the definitive 'tariff' system, in which discharge would not be considered despite the clinical condition before the stated number of years had been served. Power was seated firmly on the wards and responded little to the 'higher' management influences, which led to a degree of ward insularity and a management much distanced from their appreciation.

The major thrust of reorganization revolved around the Griffiths' style organization, which incorporates identifiable individuals for decision-making and accountability, from a Unit General Manager, through Heads of Departments, to line structures in decreasing hierarchy. A major problem to its implementation has been the staffing structure at the charge nurse/sister level with two to three nurses of this grade sharing total responsibility according to their shifts on duty. Clearly, this is antithetical to the Griffiths' decision-making and accountability concepts, as each nursing shift under a different nurse in charge is operationally different.

The objective of implementing Griffiths was to establish these levels of power from a top-down approach. However, as the real power was historically at the ward level, the implementation of the Griffiths' structure on its way down through the system, appeared to falter at the charge nurse tier. Thus the 'power' going downwards met a resisting, established 'power base', resulting in a series of conflicts. To counteract this, the Special Hospitals appeared to invert Griffiths and erode the power base of the wards away from the charge nurse grade by basing 'clinical specialists' and ward managers closer to the wards themselves, which is likely to result in a 'smoothing out' of power structures and prise open the wards to external influences.

The new managerialism appears to be based on a 'nationalization' process, that is, to structure the Special Hospitals' organization alongside their contemporaries in the National Health Service and to make them responsive to managerial considerations. However, for this to be convincing, any emerging managerial assembly will need to appreciate the differences in cultures between the Specials and the NHS. Preceding management structures over the decades have failed to create the levels of change they may have desired due to their failure either to confront the cultural component or have refused to accept their existence.

Recent emphases within the Special Hospitals have endorsed rationalistic principles revolving around cost-effective care delivery, budgetary controls and income generation which has focused attention on the 'nationalization' of the system, and raised the level of resistance from the Unions. With the traditional union closed-shop approach within the Specials, management related only to the POA in negotiating changes. However, as other representations have been developed namely, COHSE and the RCN, the gulf between the POA and management has become entrenched in conflicts which are becoming increasingly bitter, as the POA feel their power base being eroded. It can be argued that inter-union rivalry can split the cohesiveness of nursing teams on each ward. Withdrawal of labour or work-to-rule by the POA affects patient routines and care. Some patients can empower themselves by manipulating the uncertainties now engendered by inter-staff rivalries. The traditional methods used to deal with 'dangerousness' can no longer be sustained.

Quo vadis?

The two main positions in the debate appear to ensconce themselves in either the 'abolitionist' camp or the 'reformist' quarter, with both scenarios having fundamental effects on forensic psychiatry.

Abolition

Some have suggested the complete closure of the Special Hospitals, arguing that they are unsuccessful in their fundamental *raison d'être*. Firstly, in terms of their inability effectively to treat mentally abnormal offenders, it is questioned whether they can cure those illnesses that have been related, a priori, to a criminal offence (the causal relationship between illness and crime is too large a topic to enter into in this article). Secondly, they argue for closure on the grounds that the prognostic and predictive components of their function is inadequate, that is can the system give assurances as to re-offence and 'dangerousness' prediction? Thirdly, as a conclusion to the foregoing, they contend that if the Specials cannot effectively treat offenders and predict dangerousness to some degree of accuracy then the system is not cost-effective and should be dismantled. Finally, closure is supported on the grounds of isolation, in terms of the geographical distances that can exist between in-patients and their relatives, and should be replaced by smaller and more local facilities.

However, the consequences of closure of the Special Hospital system would have extensive implications for the social infrastructure which needs careful consideration. Should the offenders coming to the courts be referred towards the medical system, then both the Regional and Interim Secure Units would quickly become congested and ultimately blocked with chronically dangerous offenders, and the pressure to build more units would thus become widespread. Health Authorities would find themselves being coerced to provide beds which would inevitably result in a return to locked wards in psychiatric hospitals. The provision of halfway-houses on a relatively large scale would be necessary which is not only expensive to operate but also causes antagonism to the local communities.

For those not placed in the medical framework would by necessity be referred to the criminal system. The prison hospitals would require considerable development to cope with the influx of mentally abnormal offenders, that is, if any attempt was going to be made to alleviate their psychiatric symptoms rather than resorting to mere 'control' in the main prisons. Legislation may be necessary in order compulsorily to treat offenders in the prisons, which has wide and serious ramifications regarding lines of demarcation between crime, behaviour and illness, which is catered for in the 'Specials' by allegiance to 'medicine' and is not to the same degree in the prison service.

Reform

This group recognizes certain deficiencies in the Special Hospital system and tends towards restructure, reorganization and change. With the inception of the new Special Hospitals Service Authority, strategies for the future are being developed and, although we are not yet privy to them, it is reasonable to assume that reform is included. The reformist argument revolves around the notion of the mentally ill being mis-placed within the prison system and that other structures are both sparse and inappropriate. They suggest that it is immoral not to provide treatment to the mentally abnormal offender, and regard the Special Hospitals as the most suitable structure, albeit with varying degrees of reform. They argue that only by developing the system can an increase in knowledge and expertise be cultivated, which will lead to an expansion of treatment efficacy and predictability.

The effects of reform is, of course, dependent on what exactly they reform into, but it is reasoned that an increased medicalization of criminology and an expansion of forensic psychiatry is one likely development. Reform may have the effect of restructuring the internal organization to lessen the felt isolation of the ward staff and reduce the insularity which has allowed sub-culture formation. The production of hospital-wise coordination may construct an interdisciplinary cohesion which may form the basis for addressing the hostile public opinion, reluctance of psychiatric units and Regional Secure Units within the National Health Service to accept 'stigmatized' Special Hospital patients.

Dick et al (1989) proffered four models for reform:

(i) the island model in which the three Specials remain independent of each other, develop services in each of their own areas and cultivate relationships with organizations in their locale;
(ii) the pyramid model, where the Special Hospitals are at the apex and serve Regional Services and Regional Secure Units below them, with consultants and staff sharing responsibility at all levels through the system;
(iii) the local service model, which disperses the functions of the Specials into a number of smaller units, perhaps at Regional Health Authority level;
(iv) the territorial model, in which the Specials lose their 'Special' tag and become a part of a network of services benefiting a defined community, operating a psychiatric morbidity role for a given population.

The future of the Special Hospital is woven with many macro issues – penal and mental health reform in general, the strategic position of psychiatry and citizenship and civil rights. Not least, there is the semiotic representation of dangerousness, madness and evil. Current thought has reinforced this nineteenth century trinity. There are the immediate dangers to the planet from ozone depletion and hidden pollutants from uncontrollable technology. There is also the new fascination with the serial killers of the Hannibal Lecter type. (Even the psychiatrist is capable of transcending ethical limits of 'human exploration'.) Thus moral panics focused on a select band of individuals as the epitomization of absolute evil and unleashed dangerousness create the comforting aura that society has logical continuity. The Special Hospital, then, remains the bastion of society's nightmares.

Notes

The views of the authors are personal ones and in no way reflect those of the Special Hospitals.

References

Coldwell, J.B. and Naismith, L.J. (1989), 'Violent incidents on special care wards in a Special Hospital', *Medicine, Science and Law*, **29**:116–23.

DHSS/Home Office (1986), *Offenders Suffering from Psychopathic Disorder*, London: Joint DHSS/Home Office Consultation Document.

Dick, D.H., Deardon, R.W., Foley, S. and Gardner, J.C. (1989), 'Prejudice and pride: a report about Rampton Hospital ten years after the Boynton Report', presented to the Right Honourable Roger Freeman MP, Parliamentary Under Secretary for Health, January 1990.

Evans-Pritchard, E.E. (1937), *Witchcraft Among the Azande*, Oxford: Oxford University Press.

Foucault, M. (1978), 'About the concept of the dangerous individual in nineteenth century legal psychiatry', *International Journal of Law and Psychiatry*, **1**:1–18.

Gunn, J. (1985), Foreword to Webster, C.D., Ben-Aron, M.H. and Hucker, S.J. (eds), *Dangerousness, Probability and Prediction, Psychiatry and Public Policy*, Cambridge: Cambridge University Press.

Hamilton, J.R. and Bullard, J. (1990), 'Dangerousness: which patients should we worry about', in Hawton, K. and Cowen, P. (eds); *Dilemmas and Difficulties in the Management of Psychiatric Patients*, Oxford: Oxford University Press.

HMSO (1980), *Report of the Review of Rampton Hospital*, Cmnd 8073.

Kerr, C. and Siegel, A. (1954), 'The interindustry propensity to strike – an international comparison', in Kornhauser, A. et al (eds), *Industrial Conflict*, McGraw-Hill.

Mason, T. and Chandley, M. (1990), 'Nursing models in a special hospital: a critical analysis', *Journal of Advanced Nursing*, **15**:667–73.

Mental Health Act (1983), Section 118: *Draft Code of Practice, Mental Health Division*, Department of Health and Social Security, Alexander Fleming House, Elephant and Castle, London.

National Board for England and Wales (1982), *Syllabus of Training: Professional Register – Part 3*, Registered Mental Nurse, ENB, London.

National Health Advisory Service, DHSS, Social Services Inspectorate (July 1988), 'Report on services provided by Broadmoor Hospital'.

National Health Service Training Authority (1990), National Occupational Standard, Care Sector Consortium, Health Care Support Worker Project.

Richman, J. (1989), 'Psychiatric ward cultures revisited, implications for treatment regimes', paper presented to the British Sociological Association Annual Conference, Plymouth.

Royal Medico-psychological Association (1968), 'Memorandum on the Special Hospitals', evidence to Sub-Committee B of the Estimates Committee.

Scott, P.D. (1977), 'Assessing dangerousness in criminals', *British Journal of Psychiatry*, **131**:127–42.

11 The social relations of HIV testing technology

Evan Willis

As the AIDS epidemic has progressed, few of the many issues that have been raised have been as controversial as the now highly politicized issue of HIV testing. Debates over when and where testing should occur and who should be tested have been a major part of the unfolding societal response to the epidemic. This relatively simple tool for detecting the presence of antibodies to the HIV virus in the blood is the first evidence of infection and, in terms of the current state of knowledge, the first step in the patient career trajectory that for most, if not all, carriers will progress over a period of time to full blown AIDS and death.

This chapter focuses on the social relations of HIV testing technology. The technology has been defined by the World Health Organisation as

> a serologic procedure for HIV antibody (or antigen) for an individual person, whether recommended by a health care provider or requested by the individual (WHO 1987:7).

The specifically sociological analysis of the social relations of medical technologies has a different focus from other approaches to assessing medical technologies; from the epidemiologist's randomized control trials to evaluate effectiveness, to the health economist's cost benefit analyses and the philosopher's bioethical considerations. It is argued that assessing the social relations of particular medical technologies is an essential part of the overall developing field of medical technology assessment.

Applying such an analysis to the technology of HIV testing, this chapter aims to clarify the complex issues relating to HIV testing; in particular to analyse the complex set of social processes including the power relations and ideological context in which the test is located. It attempts to provide a sociological analysis of the test, not couched only within the AIDS debates, but in the broader debates about medical technologies in general. Few examples of medical technologies developed in recent years have had such a dramatic impact as that of HIV testing technologies; it is enmeshed in a very complex social context and the social relations present an important case study of the place of technologies in general, and medical technologies in particular, in the conduct of human affairs.

One point of clarification about the sociological analysis should be made before beginning. The chapter makes the assumption that different sorts of knowledge such as scientific knowledge about AIDS is socially constructed in its social, cultural and historical context. As Latour and Wolgar (1986:242) argue 'Scientific activity is not "about nature", it is a fierce fight to construct reality' (see also Treichler 1990).

Since the identification of the disease in the early 1980s, AIDS has come to represent a major new challenge with profound sociological and social policy implications. The dominant paradigm of knowledge about illness and treatment which had come to be known as Western Scientific Medical knowledge promoted a bio-medical paradigm of disease – that disease was an individual and biological phenomenon that could be alleviated by 'technological fixes' mainly in the form of chemotherapeutic technology (Willis 1989)

Armed with this biomedical paradigm, medical researchers have concentrated on the search for a 'technological fix' of the chemotherapeutic kind in the form of a vaccine or other treatment. Thus far, however, the search for a 'magic bullet' in the form of vaccine or other treatment has been unsuccessful. Furthermore, the history of the treatment of infectious diseases in general, and sexually transmitted diseases in particular (Brandt 1988), suggests that it is unlikely that development of such treatments will easily end the epidemic. The 'sense of problem' thrown up by the biomedical paradigm focuses on a technological fix, this time with a change of emphasis from chemotherapeutic to other medical technologies; in this case the screening technology of testing.

The strategy of prevention by testing exists alongside, and at times in contradiction with, the 'sense of problem' thrown up by a more social model of health and disease causation; that is widely disseminated public health education campaigns on the causes of the disease syndrome and the practical means of preventing its transmission. The two strategies rest upon different foundations. The strategy of prevention by education rests upon viewing the disease as a social phenomenon, the testing strategy rests upon seeing it as more of an individual and biological phenomenon. Both strategies contain assumptions, on which they rely for their effectiveness, in particular on the intervention (either testing or education) having consequences in less risky subsequent behaviour. The education strategy rests upon the assumption that people will act in terms of their 'new' knowledge about the threat to their health, while the testing strategy rests heavily on the assumption that knowledge of antibody serological status will affect subsequent behaviour. Both these are empirical issues on which there is now a considerable body of evidence to be reviewed later.

Of course it might be argued that both are relevant but they come into conflict in the context of competition for scarce financial resources. Should finite health budgets allocated to contain the spread of AIDS be put into mass education campaigns or into establishing testing facilities? This decision is particularly acute for Third World countries where, in a number of cases, the cost of a single HIV test exceeds the annual per capita spending on public health (Ankrah 1989:271). Furthermore, the different experience of the AIDS epidemic in the developed, as opposed to the developing world, has been important to the way that knowledge about the epidemic has been constructed. The social construction of the definition of AIDS cases by the World Health Organisation has, since the availability of HIV testing technology, been based on the presumed link between antibody positive serological status and the eventual development of ARCs (Aids Related Conditions). Epidemiological surveillance of the progress of the epidemic was based on this definition. In developing countries, however, the impracticality on cost grounds of HIV testing being available even to high risk groups, led to a recent change in the definition of AIDS cases to the presentation of an ARC alone, without reference to HIV serological status (Colebunders and Heyward 1990).

Refocusing upstream

The differences between the two approaches are illustrated by the well known analogy in the sociology of health literature, developed by John McKinlay (1981) to analyse health interventions. The practice of health care based on the biomedical paradigm of medical knowledge, he argues, is too focused upon rescuing people from the 'river' of ill health. Instead he argues for the need for 'refocusing upstream – to see who is pushing them in, in the first place!' While, of course, both interventions are relevant, emphasis upon the strategy of utilizing limited financial resources on testing as a response to the epidemic of AIDS may be considered a 'downstream' approach. Putting more emphasis in resource allocation terms on preventing people becoming antibody positive in the first place is the more 'upstream' approach thrown up by a more social paradigm of ill health.

The assumption that risk reduction for the population as a whole can be based on knowledge of individual antibody status focuses the responsibility for change downstream to the individual level rather than upstream to focus on changes in normative sexual behaviour for the sexually active population as a whole. Testing as a downstream strategy for controlling the epidemic is more consistent with existing social and political structures of power relations than the modification to group behaviour implied by focusing upstream. As a technological fix, therefore, in dealing with the epidemic, testing is less threatening to the existing social order. Not only are power structures based on class and patriarchy less challenged by testing, but it also provides a substantial avenue for capital accumulation in the marketing of testing technology which emphasis upstream does not.

The development of testing technology

The original 'sense of problem', then, which gave rise to the search for a test to detect the HIV virus, was the question of identifying those individuals who would eventually develop AIDS and therefore may have the potential to infect others with the same virus. This was particularly the case with the need to protect and ensure the integrity and legitimacy of the supply of whole blood, blood products and tissue, routinely used in the everyday practice of health care by identifying those biological products that were potentially infectious to others. Before the test was available, a considerable number of people the world over contracted AIDS through the routine transfusion of blood, or as haemophiliacs dependent on blood products.

A number of different biochemical tests have been developed in response to this urgent need. These will detect the presence of antibodies to the HIV virus with varying degrees of success. There is not a test it should be noted, but product diversification with competing brands marketed by different diagnostics manufacturers.

The first HIV test was licensed by the United States Federal Drug Administration in early 1985; stressing that the test was for screening blood products rather than diagnostic purposes. Other tests have subsequently been and are being developed but are less widely used for a variety of purposes. More recently, tests have become available on a do-it-yourself basis (in one, place a blood drop on a reagent strip, in a short time, a plus or minus appears). The advantages of these tests it is claimed are that not only are they cheaper and quicker to produce a result, but they remove the requirement of skilled laboratory technicians to 'read' the assay. This is presented as making testing more objective by removing the subjectivity of the technician in interpreting the result; an accurate result being possible with one brand, even if the user is colour blind! (Kelly 1990).

Assessing testing technology

In assessing the testing technology, several difficulties should be considered, which occur with the existing testing technology. These have collectively imposed limitations on the overall usefulness of the test in controlling the epidemic; that is with the uses to which the test is put. Firstly, while the conventional technology assessment criterion of safety is not at issue, there are questions of the effectiveness and accuracy of the tests; usually expressed by the epidemiological criteria of specificity and sensitivity. *Specificity* refers to the ability of the test accurately to identify those who are antibody negative, that is have not been infected by the HIV virus, the presence of antibodies which are detected by the test. A highly specific test is one which minimizes the number of false-positives; that is people who are not antibody positive but whom the tests indicate are. *Sensitivity* refers to the ability of the test to minimize false-negatives; people who are antibody positive but whom the test identifies them as negative.

Over time, the sensitivity and specificity of the tests have been refined to be comparable or better than similar biochemical tests. The early ELISA (Enzyme Linked Immunosorbent Assay) tests had problems of specificity, showing a high level of false positives, particularly amongst women who had borne children (Goldwater 1986). The original use of the test to screen the blood supply meant that *sensitivity* was a more important criterion than *specificity* and the test developed accordingly. From the objective of screening the blood supply it was more important to ensure that the blood of donors, who were in reality antibody positive, was not used, than it was for there to be some false positives, since these could be rectified by subsequent or confirmatory testing. It should be noted that attempts to screen the blood supply have been successful. In the Auckland region of New Zealand for instance, only ten donations of blood infected with the HIV virus have been detected since screening began in 1985, with 75 000 donations processed annually (Du Chateau 1990:77). Similar figures are available throughout the developed world.

Secondly it should be explicitly noted that it is *not* a test for AIDS. AIDS is not a virus but a syndrome of conditions. Thus, there is no such thing as an AIDS test. Strictly speaking, therefore, an HIV test is not a diagnostic test, since it does not detect the presence of a disease as such but the presence of antibodies to a virus (and by derivation, the presence of the virus itself) which at some future time will make the individual carrying it susceptible to the range of opportunistic infections. There has frequently been a conflation of the two issues in public discussions.

Thirdly, there is the serious issue of the 'window of infection'. Antibodies to the virus are not measurable for a period of time after the actual transmission of the HIV virus. During this 'window' of between three to six months, the person may test antibody negative when they are able to transmit the virus to others. This is a limitation in the usefulness of the test in controlling the epidemic. It means what must be said is that the person tested antibody negative on the day of the test itself may subsequently develop antibodies to the virus. In screening the blood supply, for instance, this difficulty is coped with by the 'lifestyle' questionnaire which seeks to identify high risk sexual practices, particularly amongst homosexuals. High risk sexual practices amongst multi-partnered heterosexuals are less easily identified.

Limitations exist in the ability of the test to 'certify' a status of antibody negative. The Centre for Disease Control in Atlanta, Georgia, one of the leading world research institutions on AIDS, defines and certifies antibody negative status as testing antibody negative, six months after stopping defined high risk behaviours such as unprotected anal intercourse (Posnick 1990). In the earlier stage of the epidemic, some entrepreneurial attempts were made in the United States to deal with the threat of infection in the 'sexual marketplace', not by the encouragement of safe sex practices but by the establishment of 'AIDS free dating services'. These operated on the basis of a certificate of antibody negative status which could be shown to potential sexual partners, but, quite apart from the possibility of forgery, they could not guarantee a clean bill of health. In more recent times, a code for communicating serological status in the language of seduction has evolved; telling someone that you donate blood regularly is a means of establishing 'safety' in the awkwardness of negotiating around the issue of AIDS!

HIV testing as screening technology

What started as a means for screening the blood supply rapidly came to have wider screening applications. Here the social relations of HIV testing technology became complex. Screening technologies in general detect asymptomatic signs which are held to be predictive of future symptoms and illness, unless intervention occurs. There are substantial differences between screening for the presence of antibodies to the HIV virus compared to blood pressure, cholesterol or cervical screening. Firstly, the consequences of a positive antibody test are rather different to other screening tests. Relatively simple treatment is available for abnormal blood pressure, cholesterol or cervical smears but not for positive antibody status. In this sense, it is closer to diagnostic tests for cancer than to screening procedures. No cure exists, only in very recent times with increasing evidence of the effectiveness of drugs such as zidovudine (AZT) and pentamidine isethionate used either prophylactically or in treating ARCs, can something be done in response to a positive test.

The social relations of the test are therefore much more problematic, because of the adverse psychosocial effects that can result Counselling both before and after the test is now considered essential both to maximize the ability to affect behaviour and also to minimize the adverse effects of a positive test. It is for this reason that do-it-yourself tests have been restricted in availability in some countries like the United States and banned from use in others like Australia and New Zealand. Counselling serves to provide a framework for interpreting the test result and is universally held to be essential, especially if it should be positive. ('It's not a death sentence' etc) Interpretation of the meaning of a negative result is also necessary ('it's not a carte blanche to do as you want' etc). It is by now well established that the considerable psychosocial consequences that can follow from disclosure of a positive test result include severe anguish, fear, depression and suicide Perry et al 1990). Indeed fear of the consequences of a positive test result has been frequently found to be an important reason why a significant proportion of people who practise high risk activities do not present themselves for testing, even when specifically invited. In the study by Dawson et al (1990a:11), 43 per cent of the 502 homosexual men in the United Kingdom who had not been tested reported the reason as being concern about the effects of the test on their psychological well being.

Secondly, and more importantly in sociological terms, the meanings attached to being antibody positive are unmatched by any other screening test. All medical screening tests confer both a social as well as a biological status, but being found to have elevated blood cholesterol levels doesn't have the same meaning because it does not mean a status transformation into a deviant social identity. Being labelled antibody positive has come to have much greater implications for the person's identity than, say, being labelled hypertensive, akin to becoming a master status. The lengths to which this may be carried can be seen from the proposal from the American New Right writer William F. Buckley who proposed tattooing on the upper forearm and buttocks to warn others! (*New York Times* 18 March 1986:A27).

The social consequences of being labelled seropositive (or even sometimes participating in screening programmes at all) have been profound. In instances catalogued by the WHO (1987), testing has opened the way for all manner of discrimination, such as social isolation through loss of friends, severe economic loss, cancellation of insurance, and restriction of opportunities for employment, schooling, housing, health care and social services, including loss of jobs, accommodation, friends etc. The HIV test therefore can never be considered 'just another screening test'. Because of this, issues of privacy, confidentiality of test results and informed consent in taking the test are crucial aspects of the social relations of the testing technology in attempting to prevent the negative implications of the societal reaction to the labelling process.

These social processes operate in considerable part by the ideology of victim-blaming, common in many areas of society, attributing social structural conditions to human agency. Since the beginning of the epidemic, people who are HIV positive have been labelled 'victims' and a label attached reflective of the mode of transmission; either an 'innocent victim' as in the case of haemophiliacs or children, or by usually unstated implication, a 'guilty victim' by deviant forms of behaviour such as homosexual relations or drug taking. No other screening tests carry such ideological baggage; discussion is never heard of an innocent victim of elevated blood cholesterol or the presence of dysplasia on the cervix. The historical equivalent would be if a screening test had been available for leprosy, or with the development of the Wasserman test for syphilis (Fleck 1979). As McCombie (1986:455) argues, contagion has a social definition not only in 'primitive' societies studied by medical anthropologists, but also in the context of western scientific medicine. The rapid diffusion and implementation of HIV testing technology, he argues, demonstrates how cultural beliefs and attitudes affect both medical practice and public health. Arguably, all people who are antibody positive are victims, but the attachment of anthropological meanings of pure and impure, taboo violation and pollution are powerful social meanings to the conditions which even researchers in the field find it difficult to escape from.

HIV testing as social control

When considering the uses to which the test has been put beyond screening of the blood supply, it is clear that it is designed and used as a means of social control. Legitimated by the ideology of victim blaming, it is a device for detecting social deviance more than screening to permit intervention in the interests of individual or collective health. In a number of uses to which the test is put, its ideological basis is more akin to testing for drugs in sport or blood alcohol levels in driving. Here the test is being used for punitive rather than preventive purposes. In instances such as insurance and immigration, the individuals are required compulsorily to take the test in order to secure access to some desired end such as a job or superannuation. In the instance of tests for drug taking in sport or drink driving, there is compulsion to take the test and sanctions if social deviance is detected; severe legal ones in the case of 'excessive' blood alcohol readings while driving. Sanctions are also attached to antibody positive testing; denial of access to desired ends such as migration, life insurance cover etc.

In these forms of punitive testing, furthermore, the definition of being HIV positive is frequently differently socially constructed from when the test is used for preventive purposes. When screening applicants for the American Armed Forces, for instance, if applicants test antibody positive on an initial blood sample, they are called in to provide a second sample for purposes of verification. If they decline to appear, however, they are defined antibody positive and added to the epidemiological incidence rate (Burke et al 1990:2074). Although the British Medical Association has a policy that screening before surgery is unnecessary, two London teaching hospitals have a policy of asking patients to identify any risk behaviours they may have engaged in. If they admit to any, they are requested to be HIV tested.

If they decline, they are defined as infected and extra precautions taken in operating theatres (*The Guardian*, 4 September 90:2). Clearly the definition of seropositive status has a social component.

So, a test which started out as a health screening test has increasingly come to be used as an instrument of surveillance and social control of sections of the population. Historically, we know from Foucault (1979) that sexuality is not a new area of surveillance but has been a part of the construction of social identity and social policy since the nineteenth century, serving, as Weeks (1989:9) points out, as a barometer of social anxiety and a conductor of social tensions. The development of a technological means to detect deviant behaviour in this arguably represents both a continuity and a new phase in the political surveillance, both of individual bodies and of populations. Designed and developed originally to assure the safety of the blood supply, the ends to which it has subsequently been increasingly utilized are those of social control; of determining and ultimately certifying the boundary between deviance and normality. Deviance not only in the sense of detecting traditionally proscribed practices such as homosexuality and drug taking, but also in the more general Parsonian sense of illness itself as deviance. But whereas Parsons analyses most illness as conditionally legitimated deviance, being labelled as HIV positive is in effect conditionally illegitimate unless an 'innocent' mode of transmission can be established. The certification of serological status can be analysed sociologically as part of the general societal process towards the medicalization of deviance (Conrad and Schneider 1980). This is particularly the case where notification to state authorities is compulsory.

Certification is also directed towards protecting the profits of life insurance and superannuation companies, restraining government spending in meeting the cost of medical treatment for immigrants or refugees who are HIV antibody positive, as well as providing an arena for capital accumulation for pharmaceutical and diagnostics manufacturing companies in marketing the tests. Such attempts at control through certification of course have the same flaw as the 'AIDS-free dating services' which proliferated in the early stages of the epidemic in the United States.

Attempts at control however also elicit responses of resistance or contestation by those towards whom the control is directed. The social relations of testing technology are contested as are the social relations of AIDS as a whole, by the committed and articulate gay writers and activists who have struggled to challenge the threatened hegemony of such interpretations in both their writings and their involvement in setting public policy. This has been achieved more successfully in some countries than others. Contestation over the social meaning of test results and AIDS as a whole continues, in particular over the discourse of AIDS; 'Victims' become simply 'People with AIDS' and so on.

The test in sociological perspective

Implicit in the discussion thus far are a number of central sociological issues to be briefly discussed here. Understanding the social relations of HIV testing is illuminated by reference to a very basic sociological and social issue: that of the relationship between the individual and society. The centrality of this issue has been detailed in particular by C. Wright Mills (1970) in his analysis of the relationship between private troubles and public issues; between biography and history or between agency and structure. This issue underlies the debate about the extent to which testing should be compulsory or voluntary; about whose interests are served by individuals taking the test and about how the epidemic might best be minimized.

The debate can be illustrated by considering how the parameters of the test have been defined and interpreted. What reading of the results of the biochemical assay would be constructed as defining the presence of antibodies to the HIV virus? From a societal viewpoint, especially in screening the blood supply, sensitivity is a more important criteria than specificity. It was considered more important to be overly sensitive so that no false negatives make it into the blood supply; erring on the side of caution for safety's sake (also to avoid litigation from haemophiliacs). From the individual point of view however, specificity is arguably a more important epidemiological criterion than sensitivity; the potential consequences for the individual of showing a false positive are likely to be worse than a false negative. As outlined above, these consequences will be severe anguish, and may include denied applications for various desired ends, aborted pregnancies and even suicide.

The policy representation of this classic sociological question has been cast as a tension over the extent to which the individual rights and liberties should be maintained consistent with pursuing the general aim of containing the epidemic. A variety of answers to this question have arisen, mainly organized around the issue of the extent to which compulsion is justified with appeal to some notion like 'the common good.' The issue of compulsion applies not only, it should be noted, actually to taking the test, but also to making the results of the test publicly known. That the test has a private purpose has been clear: to inform the person taking the test of their serological status. It is the extent of the public purposes to which the test result may be put that has been a contentious issue especially around the extent of confidentiality of test results.

In most developed countries, advocates of voluntary testing as the best strategy for controlling the epidemic have narrowly held the line against compulsion. The argument has been that only voluntary testing works. It has been argued that making it compulsory and recording identifying information with it, to utilize the results for the purposes of social control, will only drive underground those who are most at risk. In the specific cultural context of the United States, forms of compulsion have been introduced with varying results. The general policy goal, articulated by the WHO (1987) has been to minimize the individual effects while at the same time aiming to maximize the societal public health effects. How the test actually operates, however, involves considering the social context in which the test is located.

Testing as social process

The sociological analysis of medical technologies in general and HIV testing in particular, with its emphasis on the social relations of the technology, analyses the specific technological intervention, not narrowly as an event, but broadly as part of a social process. The social context is constituted by a social process and has both a lead up and an aftermath, to be considered here briefly in relation to HIV testing. It involves considering such aspects as the pathway to the test, the process of disclosure of test results and its effect on the individual as well as the uses to which the test results are put.

The pathway to taking the test is often complex and varies from country to country. It enables consideration of the question *Why test*? As Weiss and Thier (1988) ask 'if testing is the answer, what is the question?' They argue that 'it is the fit between the purpose of the test and the proposed test's attributes that determines whether or not to go ahead and test' (1988:1011). A distinction might be drawn in the reasons for testing which exist in different parts of the world, between voluntary and involuntary pathways. Where taking the test is compulsory, it usually has a gatekeeping function. The person being tested is required to have certification of negative antibody status, either as a means of gaining access to some desired end such as life insurance, pre-employment, premarital or pre-immigration, screening. Alternatively, it might be to allow others in the general occupational category of 'bodyworkers' to assess the potential threats to their occupational health and safety as in the instances of premedical and dental treatment, and prisons.

Where the pathway to taking the test is voluntary, the process is more complex. As indicated previously, the assumption on which testing lies is that knowledge of serological status will result in behaviour change in the direction of less risky behaviour. The rationale for voluntary testing and being informed of the result has been that it will influence transmission-related behaviours. This assumption has often been accepted uncritically but as Fitzpatrick et al (1989) argue, the question of whether the test works as a prevention strategy is a deceptively simple one with caution needed in interpreting data from studies on the effects of testing.

If voluntary testing is directed at individual behaviour change, then testing for epidemiological surveillance purposes is directed generally in determining the course of the epidemic. Here again, however, are complex issues of voluntarism and compulsion. Does the general societal 'need' to chart the progress of the epidemic outweigh the interests of individuals to voluntarily testing and keeping confidential the results? In most countries it is accepted that the need for epidemiological surveillance exists and that some degree of intrusion onto individual rights is acceptable, often with individual safeguards such as anonymity of samples, so the identity of positive readings cannot be established. One implication of this practice is for the accuracy of epidemiological evidence on incidence since individuals may be tested more than once.

Public health legislation in most countries provides for mandatory notification of certain diseases to State public health authorities, developed historically to deal with the threat posed to social order by highly contagious diseases. Despite AIDS not being a contagious disease, the framework provided by this legislation has been utilized to formulate a State response to the epidemic. What is notifiable, however, varies in different parts of the world. In both Australia and New Zealand, notification of an ARC is required. In July 1989, twenty-eight of the United States of America required notification of persons who test HIV positive, with similar proposals before the legislatures of many other states of the union (Gostin, 1990:1962). In Canada, eight provinces have mandatory reporting of HIV infection (AIDS Newsletter 1990:7).

The use of mandatory screening for epidemiological purposes has been advocated, but as Weiss and Thier (1988) argue, to be justified it must be demonstrated that the effect in slowing the epidemiology of the disease is superior to the effect of interventions possible without testing. Making mass testing compulsory is likely to have the effect of driving the disease underground, particularly if identification is attached. Concerns about the confidentiality of the results may lead to some people who engage in high risk activities declining to be tested for these reasons. Amongst the US states which have legislated to refuse confidentiality to those who test HIV positive, is the state of Colorado, where identification requirements for people taking HIV tests saw applications for such tests drop 600 per cent in three months (Watney 1987:44).

Conclusion

Technology assessment involves considering questions not only about the impact which a particular test will have, but the whole social context in which that test is located. To focus on both involves considering the social relations of a technology; as the example of HIV testing analysed here makes patently obvious.

Analysing the social relations of the technology has meant considering the relationship between the individual and society; a central issue of the social sciences in general, and sociology in particular. Much of the debate, both at the scientific as well as the lay level, has cast this relationship in relation to HIV testing as a contradiction; a zero-sum relation in which protecting the community from infection is incompatible with protecting the rights of HIV infected people.

In this chapter, I have argued that the contradiction does not exist. The spread of HIV infection will only be helped and not hindered by the protection of these individual rights, since to mandate for widespread punitive rather than educative testing will only drive those who may be HIV positive underground.

Because of the problematic and complex social relations of the test, people who engage in behaviour highly risky of HIV transmission, have either availed themselves of the opportunity to be tested less than public health policy makers have hoped, or have not modified their behaviour in the light of the test results in the straightforward way predicted by a biomedical model of disease prevention and control. A complex range of social factors influence the decision to take the test and how to respond to the result. Sexual identity is one of those factors; for example Kippax et al (1990:27) found in Sydney that those men with a comfortable gay identity, indicated by strong attachment to the gay community, and having disclosed their identity were most likely to have been tested and to have changed their behaviour, irrespective of the outcome of the test.

At the same time, a powerful technological imperative towards testing gathers momentum. Testing is moving from the unusual to be part of the routine of 'normal science', like other screening tests such as blood pressure. This is being reflected at state level where, as in New Zealand, the department of Health has just established a benefit to be paid for HIV testing. This is also being spurred by developments in the technology such as the availability of do-it-yourself tests. Rapid results will most likely mean growing use in settings such as health care. It may also mean a form of empowerment with gay and other organizations being able to conduct their own testing within the context of counselling.

The technological imperative surrounding the test can be understood, in Foucault's terms, as an expression of power which represents the potential to be either productive or repressive. In the social context surrounding testing, whether the effects for the individual concerned is repressive or liberating, depends very much on the attention paid to the social relations of the technology.

References

AIDS Newsletter (1990), Bureau of Hygiene and Tropical Diseases, September, London.

Ankrah, E.M. (1989), 'Aids: methodological problems in studying its prevention and spread', *Social Science and Medicine*, **29** (3): 265–76.

Brandt, A. (1988), 'AIDS in historical perspective: four lessons from the history of sexually transmitted diseases', *American Journal of Public Health*, **78** (4): 367–71.

Burke, D., Brundage, J., Goldenbaum, M., Gardner, L., Peterson, M., Visintine, R., Redfield, R. and the Walter Reed Retrovirus Research Group (1990), 'Human immunodeficiency virus infections in teenagers: seroprevalence among applicants for US military service', *Journal of the American Medical Association*, **263** (15): 2074–7.

Campbell, C. (1990), 'Women and AIDS', *Social Science and Medicine*, **30** (4): 407–15.

Cates, W. and Handsfield, H. (1988), 'HIV counselling and testing: does it work?' (editorial), *American Journal of Public Health*, **78** (12): 1533–4.

Ciesielski, C., Bell, D., Chamberland, M., Marcus, R., Berkelman, R. and Curran, J. (1990), 'When a house officer gets AIDS', *New England Journal of Medicine*, **332**: 1156.

Coates, T., Morin, S. and McCusick, L. (1987), 'Consequences of AIDS antibody testing amongst gay men', *Abstracts of the Third International Conference on AIDS*, June, Washington, DC.

Colebunders, R. and Heyward, W. (1990), 'Surveillance of AIDS and HIV infection: opportunities and challenges', *Health Policy*, **15**: 1–11.

Conrad, P. and Schneider, J. (180), *Deviance and Medicalisation: From badness to sickness*, St Louis: Mosby.

Dawson, J., Fitzpatrick, R., McLean, J., Hart, G. and Boulton, M. (1990a), 'The HIV test and sexual behaviour in a sample of homosexually active men', paper presented to conference on 'The Social Aspects of AIDS', April, London.

Department of Community Services and Health (Australia) (1990a), Consultation Paper No. 2, Canberra: Government Printer.

Department of Community Services and Health (Australia) (1990b), Paper No. 5, Canberra: Government Printer.

Du Chateau, C. (1990), 'AIDS: Apocalypse how?' *Metro* (New Zealand), October, pp 72–82.

Ellis, C. (1990), 'HIV infection and foreign travel', letter, *British Medical Journal*, **301** (27), October, pp 984–5.

Fitzpatrick, R., Boulton, M. and Hart, G. (1989), 'Gay men's sexual behaviour in response to AIDS – insights and problems', in Aggleton, P., Hart, G. and Davies, P. (eds), *AIDS: Social Representations, Social Practices*, London: Falmer Press.

Fleck, L. (1979), *Genesis and Development of a Scientific Fact*, Chicago: University of Chicago Press.

Foucault, M. (1979), *The History of Sexuality*, Vol. 1, London: Allen Lane.

Garland, F., Mayers, D., Hickey, T., Miller, M., Shaw, E., Gorham, E., Bigbee, L. and McNally, M. (1989), 'Incidence of human immunodeficiency virus conver-

sion in the US navy, and marine corps personnel, 1986 through 1988', *Journal of the American Medical Association*, **262** (22): 3161–5.
Goldwater, P. (1986), *AIDS: The risk in New Zealand*, Auckland: Penguin.
Gostin, L. (1990), 'The Aids litigation project: a national review of court and human right decisions, Part 1: The social impact of AIDS', *Journal of the American Medical Association*, **263** (14): 1961–70.
Hindley, D. (1990), 'Should you take the test?', *The Pink Triangle* (New Zealand), September/October, pp 20–2.
International Organisation for Migration/World Health Organisation (1990), *Seminar on Migration Medicine*, February, Geneva.
Kelly, E. (1990), 'The search for a rapid HIV test', *AIDS Patient Care*, February, pp 25–9.
Kippax, S., Crawford, J., Connell, R., Dowsett, G., Watson, L., Rodden, P. and Berg, R. (1990), 'The importance of gay community in the prevention of HIV transmission', *Social Aspects of the Prevention of AIDS Project: Study A*, Sydney: Macquarie University.
Latour, B. and Wolgar, S. (1986), *Laboratory Life: The Construction of Scientific Facts*, Princeton: Princeton University Press.
Life Offices Association of New Zealand (LOA), (1990) *Annual Review*, Wellington.
Lyter, D., Valdiserri, R., Kingsley, L., Amoroso, W. and Rinaldo, C. (1987), 'The HIV antibody test: why gay and bisexual men want or do not want to know their results', *Public Health Reports*, **102** (5): 468–74.
McCombie, S.C. (1986), 'The cultural impact of the "AIDS" test: the American experience', *Social Science and Medicine*, **23** (5): 455–9.
McCusker, J., Stoppard, A., Mayer, K., Zapka, J., Morrison, C. and Saltzman, S. (1988), 'Effects of HIV antibody test knowledge on subsequent sexual behaviours in a cohort of homosexually active men', *American Journal of Public Health*, **78** (4): 462–7.
McKinlay, J. (1981), 'A case for refocussing upstream: the political economy of illness', in Conrad, P. and Kern, R. (eds), *The Sociology of Health and Illness*, New York: St Martin's Press.
Mishu, B., Schnaffer, W., Horan, J., Wood, L., Hutcheson, R. and McNabb, P. (1990), 'A surgeon with AIDS', *Journal of the American Medical Association*, **264** (4): 467–70.
Morin, S., Coates, T., Woods, W. and McCusick, L. (1987), 'AIDS antibody testing: who takes the test?', *Abstracts of the Third International Conference on AIDS*, Washington DC.
Mulleady, G., Hart, G. and Aggleton, P. (1989), 'Injecting drug use and HIV infection – intervention strategies for harm minimisation', in Aggleton, P., Hart, G. and Davies, P. (eds), *AIDS: Social Representations, Social Practices*, London: Falmer Press.
Patton, C. (1985), *Sex and Germs: The Politics of AIDS*, Boston: South End Press.
Patton, C. (1989), 'Resistance and the erotic', in Aggleton, P., Hart, G. and Davies, P. (eds), *AIDS: Social Representations, Social Practices*, London: Falmer Press.
Perry, S., Jacobsberg, L. and Fishman, B. (1990), 'Suicidal ideation and HIV testing', *Journal of the American Medical Association*, **263** (5): 679–82.

Porter, J., Cruickshank, J., Gentle, P., Robinson, R. and Gill, O. (1990), 'Management of patients treated by surgeon with HIV infection', *Lancet*, **335**: 113–14.
Posnick, L. (1990), 'Study raises concern over HIV testing', *AIDS Patient Care*, 29 February.
Queen, C. (1988), 'The politics of AIDS: a review essay', *The Insurgent Sociologist*, **14** (2): 103–25.
Reid, M., Say, P. and Gardner, S. (1988), 'Response to AIDS education measures', letter, *New Zealand Medical Journal*, 4 September, pp 586–7.
Ross, M. (1988), 'Relationship of combination of AIDS counselling and testing to safer sex and condom use in homosexual men', *Community Health Studies*, **11** (3): 322–7.
Simon Rosser, B. (1988), 'Auckland, homosexual males and AIDS prevention: a behavioral and demographic description', *Community Health Studies*, **11** (3): 328–38.
Small, N. (1988), 'AIDS and social policy', *Critical Social Policy*, **21**:9–29.
Treichler, P. (1990), 'AIDS, HIV, and the cultural construction of reality', paper prepared for symposium on 'AIDS Research: Issues for Anthropological Theory, Method and Practice', June, Estes Park, Colorado.
Turnock, B. and Kelly, C. (1989), 'Mandatory premarital testing for human immunodeficiency virus', *Journal of the American Medical Association*, **261** (23): 3415–18.
Watney, S. (1987), *Policing Desire: Pornography, AIDS, and the Media*, London: Methuen.
Weeks, J. (1989), 'AIDS: the intellectual agenda', in Aggleton, P., Hart, G. and Davies, P. (eds), *AIDS: Social Representations, Social Practices*, London: Falmer Press.
Weiss, R. and Thier, S. (1988), 'HIV testing is the answer – what's the question', *New England Journal of Medicine*, **319**:1010–12.
Willis, E. (1989), *Medical Dominance* (revised edn), Sydney: Allen and Unwin.
World Health Organisation (1987), *Draft report of the WHO meeting on criteria for HIV screening programmes*, 20–21 May, Geneva.
Wright Mills, C. (1970), *The Sociological Imagination*, Harmondsworth: Penguin Books.

12 Safety as a social value: A community approach

Helen Roberts, Susan J. Smith and Michelle Lloyd

Introduction

Scotland has an unenviable record for child accidents (Blondel et al 1985) and within Scotland, if we are to judge from pedestrian casualty data, Glasgow is a particularly dangerous city for a child to inhabit (Strathclyde Regional Council 1987a). On a national and regional level, however, our understanding of the aetiology of child accidents is at a very preliminary stage. The steep social class gradient for accidents in the United Kingdom (which is steeper, even, than the socio-economic gradient for ill-health overall) has frequently been documented, but remains poorly explained; the same is true of the substantial excess of accidents among boys (Pharoah and Alberman 1990); and if there is a distinctive sociology or geography of accidents, we have yet fully to document and account for it.

This paper describes the first phase of a study based in a Glasgow housing scheme, which aims to identify factors predisposing children to be at risk of, or protected from, accidents. Rather than addressing the more usual question of why a particular accident happened, this research explores the strategies used by parents to maintain child safety in a demonstrably unsafe environment. The chapter assesses the contribution that a community study of safety can make to identifying the kinds of health promotion, urban, housing and social policies likely to facilitate accident prevention at a local level. It explores the relevance of regarding safety as a social value by examining the ways in which a desire for safety is incorporated into the routine behaviours which structure family life.

The starting point of a great deal of British preventive work on child accidents is that children are a danger to themselves, and that parents are deficient in their safety-keeping activities. Accidents happen because children and their parents are not well-enough informed, are not properly competent, or do not have the right safety equipment. The assumption is that the poor performance of mothers and children can be modified by health visitors, health promoters, or road safety officers. From this 'traditional' perspective, the direction of information on safety and accidents is very clearly 'top down'. Less attention is given to the poor performance of car drivers, the kitchens planned by people who have clearly never had to supervise a stove and young children at the same time, or architects producing balconies with enticing gaps at child level. Moreover, where such attention has been given (Sinnot and Jackson 1987; Ranson 1987), it appears to have been inadequately used by decision makers.

In an attempt to develop an alternative model of community safety, our starting point is the commonplace observation that our society is constructed with healthy, resilient, able-bodied adults in mind. Even the most superficial enquiry is sufficient to show that the majority of environments are not child-friendly. It is a simple matter of fact that parents and, in particular, mothers are the main providers of health care for their children. For us, therefore, the puzzle is not 'Why do child accidents happen?' but rather, 'How is it, that under the most unpropitious circumstances, most parents manage to keep their children safe most of the time?'

Taking this question as a starting point, the overall aims of the project discussed in this paper are as follows:

(i) To identify factors predisposing children to be at risk of, and also protected from, accidents in the home and the wider environment in a particular community.
(ii) To investigate, and learn from, the strategies which families routinely adopt to maintain safety.
(iii) To explore the practical value (and policy spin-offs) of recognizing that safety is a social value that can be (and is) learnt in a community setting.

The project is based in a post-war housing scheme in the south of Glasgow. The scheme is bounded by a railway line and railway yard, by a river, a busy road, and an area of spectacular parkland which houses the Burrell collection. The latter is shortly to be separated from the community by a new trunk road whose siting was unsuccessfully opposed by community members at a public enquiry.

Corkerhill is a community of some 580 dwellings, with a predominantly Social Class IV/V population. The unemployment rate is high, and over one quarter of households with children are headed by a lone parent. The number of households with four or more children is also significantly higher than that for the district or the region as a whole (Strathclyde Regional Council 1987b). Social class, large family size, unemployment and single parenthood are all factors which have been identified as having an association with child accidents (Wadsworth et al 1983; Sibert 1975; Brown and Davidson 1978).

This grim statistical representation of Corkerhill is, of course, only a partial view of the community. The people of Corkerhill have been active around a number of public health issues in Glasgow, and have met with considerable success. They were, for instance, instrumental in having some of the chemicals used to treat the mould which grew in their damp houses, banned. Likewise, a number of needlestick injuries to children alerted parents to the fact that gammaglobulin was requested by the local hospital from the regional Blood Transfusion Service on a named-patient basis, which led to delays which were psychologically, even if not clinically unacceptable. The community's representations led the Director of Public Health to take steps to ensure that supplies were routinely available in the accident and emergency department of the city's hospitals. More recently, during 1990, the mothers set up an accident prevention group. The formation of this group was, effectively, a starting point for our research.

The Corkerhill parents' campaign around child accidents is community led, and they are partners in this research in more than a superficial sense. From a practical perspective, they have been involved in the planning and organization of the work; theoretically the research is founded in the experiences of the community; and, materially, the group respondents in the work described below were compensated for their time. While this is a relatively low cost project, we see no reason why community involvement should be a way of getting expertise on the cheap. Having said this, it would be unrealistic to suggest that the interests of the academics and the community coincide at every point. There are different interests at stake, and while at one level members of the community may have an intellectual curiosity, as we do, in safety as a social value, their more pressing concerns are naturally focused on immediate rather than longer term strategies for reducing accidents. Nevertheless, there is an important contrast between our work and the community action studies of the 1960s and early 1970s, in that whereas the earlier work (initiated, for instance, as part of the Community Development Projects) tended to result in a flow of initiative from the researchers into community action, our project is just the converse, with the community taking the initiative to stimulate the involvement of academic researchers.

The community has recognized the need to make alliances with decision makers, policy makers and implementers, as well as academics, if the project is to be more than parochial in the very strict meaning of the word. In practice, there are difficulties with this, and it would be unrealistic to expect otherwise. The concept of the 'public *servant*' is no longer fashionable, and decision makers may develop a top dressing of 'matiness' and consultation while important decisions are taken elsewhere. The community (and indeed the researchers) have had to recognize that the behaviour of local decision makers, politicians and professionals is itself subject to social constraints. We hope that through the project we will also be in a position to analyse and understand some of these constraints and dilemmas facing politicians and policy implementers and those involved in promoting the public health.

A number of expert groups have been drawn into the project, including the local police, the health promotion department of the health board, workers with young people and drugs and the local health council. The community group is actively trying to forge links with the district and regional councils, from whom they hope to obtain some modest funds, and, equally importantly, make the high level politician and officer links which will enable the successful elements of this project to be used and adapted elsewhere in the city. While the community is aware that neither we nor they will be able to provide a knitting pattern for safety, we do expect, when the work is done, to be in a position to provide a framework for safety at home, at play and in transit at the community level. This combination of the research project and the parents action group is one we feel may be an effective way of producing local safety data, exploring practical ways of putting our findings into practice and disseminating them to other communities in the city.

Research design

The research has three components:

(i) Group interviews with parents from different types of dwelling in the area, and with teenagers from, and professional workers within, the community. These interviews are designed, among other things, to identify the main risk factors associated with childhood accidents at home, at play and in the wider environment; and they aim to uncover the strategies which parents, and others, routinely used to keep local children safe.
(ii) A household survey to establish the prevalence of child accidents and near accidents in the community as a whole.
(iii) Case studies of successful and unsuccessful accident prevention strategies in twenty households.

This chapter is concerned with the conduct and findings of the group interviews.

Child safety: group interviews

The primary aim of the group interviews described below was to establish, and learn from, the various strategies people routinely adopt to keep children safe in what is essentially a high risk environment. To meet this aim, it was necessary first to document the perceived, and actual, causes of child accidents. These 'causes' may be related to factors in both the built and social environment, and their role and relevance can be expected to vary over space and through time. Having established the character of dangerous places and events, it will be possible to go on to assess the nature and effectiveness of routine accident prevention behaviours – the kinds of strategies that families adopt, consciously and less consciously, on a day to day basis in an effort to keep children safe. By exploring these strategies, we hope also to illustrate some of the personal and social consequences of child accidents – those that do occur, those that nearly occur, and those that might occur – on the lives of parents and children. To address these issues, the project includes a series of group interviews with people (primarily mothers) who have (or have had) child care responsibilities in Corkerhill.

Methodology

Group interviews were considered an appropriate strategy for this initial part of the work for a number of reasons. First, we are interested in safety as a *social* value. That is, as a quality of life which is defined, sought after, developed and maintained, by whole communities or societies. Individual views and practices are important, then, not simply for what they achieve for that individual and their immediate family, but also for the way they shape and, are themselves mediated by, the views of others. Group interviews are an important vehicle by which this process of interaction and negotiation can be monitored.

Group interviews are, secondly, well attuned to the spirit of this particular piece of research which not only recognizes the authenticity and value of local knowledge, but also seeks to 'democratize' the research process. Group interviews have a strength, in this respect, relative to individual interviews, since they allow the people involved to make their own generalizations about the topic at hand. This contrasts with the more conventional model in which analysts interview a range of people separately and then retreat to their offices to draw out the generalizations themselves.

Other advantages of the group interview approach are outlined by Hedges (1985). He concludes they have much to offer to research in which 'the social context is important', when understanding and insight (rather than description and prediction) are required, when the work has an 'action research' component and where the interviews are part of a process of generating new ideas. These conditions all obtain in the Corkerhill research, which is rooted in the principles of qualitative research with its aim of exploring 'the realities of everyday lives as they are experienced and explained by the people who live them' (Burgess *et al* 1988: 310).

The obvious disadvantage (from some perspectives) of group interviews is that they rely on the comments of a small, and not necessarily typical, sample of residents. They do not, therefore, provide the kinds of sample estimate that inferential statistics requires to allow us to make generalizations about population parameters. On the other hand, this part of the research is not designed to make statements about the incidence and experience of accidents and related events throughout Corkerhill: that is the task of the household census. The group interviews were concerned:

(i) to develop an inventory of accident risks (to feed into the household survey stage);
(ii) to study associations between certain kinds of risks and actual or near accidents;
(iii) to provide insight into the meaning, experience and effects of accidents.

To meet these goals, typicality is not necessarily a relevant criterion. Indeed, as Mitchell (1983) argued in his eloquent defence of the case study approach to social research, atypical, even idiosyncratic cases may provide important insights into the way a community 'normally' thinks and acts.

Other disadvantages of the group interview approach are listed by Hedges (1985). The most challenging of these concern the tendency to subsume individuals into the life of the group, absorbing minority or hesitant voices to the will of the majority or dominant voices, and encouraging individuals to play a role. These are important problems. But the point here is that group interviews have a very specific role: they are not a substitute for random samples, where generalizations about a population are required; and they are not a substitute for individual interviews where the primary aim of the research is to catch a glimpse of personal details and private lives.

Although in designing the group interviews we did not seek representativeness in a statistical sense, we did take steps to ensure that as wide a range of views as possible was included in this part of the research. Thus, five types of group were targeted for interview: One group of (predominantly) mothers was drawn from each of the three main housing types in Corkerhill: those living in pitched-roof dwellings, those in flat roof housing and those in cottage type dwellings. The distinction between housing types was chosen because the relationship between dwelling design and accident risks was one of the main concerns of the Corkerhill residents who had initiated the research. Moreover, while Corkerhill is itself a post-war estate, it was built in several phases. The pitch-roof and flat-roof developments are older, and they are all flatted accommodation. The flat-roofed dwellings are generally in the worst condition, while the more recently built cottage dwellings (houses with gardens) tend to be better quality and more sought-after.

In addition to the parents groups, we interviewed a mixed group of teenagers. Teenagers, it was felt, would be young enough to remember accidents and near accidents that they themselves had been involved in, yet would be old enough to have begun to take on child care responsibilities themselves (for instance, by acting as baby-sitters for parents and other relatives). Finally, to get a perspective from other people with a responsibility for child safety, but without the responsibility for routinely maintaining this on a day to day basis, a group of professionals was convened, comprising two health promotion officers, a fire prevention officer, a health visitor, a housing manager, a community police officer, and a road safety officer from the council's roads department.

The aim at the outset was to interview each group of six to eight people three times for between forty-five and ninety minutes (except the professionals, whose interview took the form of one meeting). This means that most of our interviews were more concerned with what Burgess et al (1988) call 'in-depth small groups' than with 'once-only groups'. Our aim here was not so much to 'enable participants to understand themselves better, and to develop or change some part of their professional or personal behaviour' (Burgess et al 1988: 312). It was, rather, to allow a certain degree of familiarity to grow up between the interviewers and interviewees: to allow a common understanding and appreciation of a range of local problems to be developed. This strategy was important: all the groups 'worked' better in the later interviews. But it was vital in our meetings with the teenagers who did not really begin to speak freely until the last session.

In contrast, in the group convened with the professional workers the gap between interviewers and interviewees was much narrower, and while the practicalities of convening the same group of professionals more than once meant that a second interview was never planned, in fact a single interview was sufficient to cover all the necessary ground. This group was also of a kind where 'the interpersonal relationships of the members are secondary to the discussion of the product' (Burgess et al 1988: 311).

The interviews with parents and teenagers were conducted with two of the three authors and took place in Corkerhill. The professionals group met at the university. The teenage and parent interviewees were recruited for us by members of the parents' safety group. This meant that (in contrast to the process of recruiting the professional representatives which we did ourselves) we had no direct control over precisely who came to the interviews. We did, nevertheless, specify broadly what kind of person was needed for each set of interviews (ideally the Corkerhill residents were to have children, or to have brought children up in the recent past while living in Corkerhill). Table 1 describes the interviews finally achieved:

Table 1 Group interviews for the Corkerhill study

	Number in group	Number of years in this community	Number of meetings
Group 1 (Flat roofs)	9	12.6	3
Group 2 (Pitched roofs)	4	21.7	2
Group 3 (Cottage type)	8	15	3
Teenagers	6	9.5	3
Professionals	7		1

Child accidents: developing a community view

Over the course of three weeks[1] the following themes were discussed with the parents' groups. First, the groups addressed the issue of what an accident is, and who is at risk. We asked people what they thought of when they heard the word accident being used both in the mass media and in local life (on the bus, in the shops and so on). These ice-breaking questions were followed by discussion of the kinds of accidents that are common in Corkerhill, and about the kinds of people that seem most vulnerable to accidents. People were then encouraged to talk in general terms about the accidents, and near-accidents they had experienced, and to give an indication of the kinds of accidents that concerned them most. This first session was designed to set the scene: the group defined its terms of reference, by identifying what might and might not be construed as an accident, and by thinking about the social as well as personal consequences of living in an essentially high-risk area.

Week Two focused more directly on where accidents happen, and on the factors thought to cause them. The session began by asking the group to make an inventory of features in their home that might put children at risk.[2] The groups talked about why these features were dangerous and made suggestions on how they could be made safer. The same line of discussion was taken when the subject moved to safety outside the home. This section ended by considering the question of why, since the environment is so dangerous, there are not more child accidents in housing estates like Corkerhill.

Finally, the groups were asked to consider what impact the risk of accidents had on personal and social life. It is already well-known that other risks – for instance the risk of victimization - can severely affect people's quality of life and restrict their full participation in society (Smith 1987). Do people experience stress, worry or anxiety about accident risks? Does this affect their behaviour or the kinds of activities their children are allowed to take part in? Do people see accidents as an inevitable hazard or as something they can take steps to control? The interviews were concluded by asking the groups what they would do to make their environment safer a) if they had unlimited funds; b) if they had only enough money to make two or three of the changes they thought were needed; and c) if they had no money, but could change the way existing services and resources were organized.

Much the same kind of approach was taken when the teenagers were interviewed. However, it rapidly became apparent that a different approach was needed. The teenagers had little to say about what we called 'accidents' or, indeed about the notion (which we have tended to favour) of safety. Accordingly, we asked them to set the agenda. 'You should ask us how we got our scars...' was the advice of one of them on how to approach the question of safety and accidents with young people.

A different agenda again was needed for the interview with professionals. The group was asked questions about the incidence, causes and prevention of accidents in an area like Corkerhill. We also asked for comments on the kinds of problems the parents' groups had encountered in attempting to keep their children safe. The meeting concluded with a discussion of priorities for tackling accident risks with different levels of resourcing.

Findings

The group interviews described above were completed between November 1990 and the end of March 1991. Key themes that emerged from our transcriptions can be discussed under five headings:

(i) Accidents are part of a wider environment of insecurity First, the people of Corkerhill who took part in the group interviews did not draw a clear distinction between accidents and other risks. The threat posed by accidents is just one element in a seamless web of insecurity associated with life on a Glasgow estate. Thus, for instance, talk about accident risks in the home very easily becomes talk about health risks. 'Your house is damp ... your house isn't safe ....' Accident risks and health hazards are both embedded in the built environment, and while it might suit urban managers to separate them out for policy purposes, for the purposes of everyday life, they are

intimately associated. As one person pointed out 'That's when accidents are more likely to happen – when you're under stress. And living in these houses is stressful enough ....' Similarly, a lengthy discussion about compensation for accidents caused by alleged council negligence drifted easily into a discussion of compensation for illnesses thought to be linked to the chemicals used to protect homes against damp, and a discussion of parental tactics in reducing risks.

> They left them (chemicals to treat mould) and she eventually went tae the newspaper – the Govan Press – because she had asked them tae come oot and move them, and they didnae bother which was quite easy for one of the kids tae go doon and open it up. She got fits off the Housing Department for going tae the press (G1 I Ls564–572).[3]

In a similar way, talk about the risks of accidents tended to be interspersed with talk about the risk of crime. For instance, in discussing the risks of some external design features, mothers' anxiety about accidents tended to melt into anxiety about crime

> ... There's the cellar doors ... I mean anybody could be hiding behind them. I mean ah'm fear't going in at night. I think they're quite frightening (G1 I3 L447).[4]

The teenagers evinced a similar tendency. The local chippie figured prominently among the places they said were unsafe on the grounds that 'There's a lot o' neds up here. There's a crowd o' guys walk aboot an' that. They think they're hard' (T1 L965). Furthermore, while talking about accidents, the teenagers frequently drifted on to talk about fights and gave graphic descriptions of specific violent acts in the area. For instance, one of them observed

> The last time ah seen one [an accident] wis ... a really bad one – it wis sometime last year ... Rangers Celtic game an' the guy got the back o' his heid kicked in. There wis a' the blood lying ... (T3 L353).

Accident prevention and crime prevention were talked about interchangeably. One accident prevention strategy was identified as 'intercoms for the closes so that you can just get in and naebody else' (G1 I3 L493).

In short, although the idea of accidents may be convenient from the point of view of analysts and professional bodies, in practice, accident prevention in communities appears to be part of a set of more general risk minimization strategies in an essentially uncertain environment.

(ii) The effects of accidents are largely unpredictable, but their causes are not. As with other sources of insecurity, the precise times and places, and

chains of events which might lead to accidents, are seen as unpredictable. Whether and when a risky environment will interact with other behaviours to produce a near accident, whether that near accident will become a real accident, and whether and what injuries will arise from that accident, are not regarded as predictable. Because they bear this out, the adults' interviews could be interpreted as suggesting that it is regarded largely as a matter of luck as to whether or not their children come to harm. However, it is only the infliction (or not), and extent, of harm that is thought to be unpredictable. Parents are acutely aware of the range of factors that can cause accidents: the element of unpredictability is not in the causal factors, but simply in whether, at a given time and in a given place, that cause will have an effect. One resident thus described how a chance sighting of a broken (and subsequently vandalized) telephone wire saved both her and her children from a potentially serious accident:

> ... and them up the close ... pulled every bit o' wire doon and wrapped it round my gate. Just by luck I was coming up the stair and seen it. People coming oot my door would'nae have seen it. It was all wrapped roon ... and the wean – he would have been choked, y'know what I mean, 'cause it was right at his level (G1 I3 L1599).

Another mother pointed out

> A nearly accident. I think you have a nearly accident every day really as far as the kids are concerned – y'know, they've nearly done this and you've stopped them, nearly went near this ... or you maybe sense that there's somethin' they're doing (G1 I3 L12–40).
>
> You hear about people talking aboot folk that are natural born mothers an' things but nane o' us ur really, and the way we learn is no so much just by bein' there as ... it's when you see them nearly daein' somethin' that you realize that that's the danger or this is the danger ... That's how you get experience, by the things they nearly dae and don't dae – sometimes the things they dae.

(iii) Child safety is as much about environment as behaviour In their discussions of causal factors, parents especially, took the view that child safety is limited primarily by issues relating to design and affordability. While social scientists will (and perhaps should) always question 'common sense' or folk explanations of why problems exist, it is hard to dispute that the following features of the Corkerhill environment – as identified by parents – constitute a threat to child safety. The flat roofs on some of the dwellings, for instance, were seen not only as the cause of the damp and mould in many flat-roofed homes, but also as a risk for children. The roofs are easily accessible and tend to be used for sunbathing in the summer. As the teenagers observed '... its easy tae get up ... if yer runnin straight forward there's no anythin' tae stop you fae fallin' (T2 L365–7). The

verandahs and stairs on the flat-roofed homes were also seen as hazardous. 'There's a space', one teenager observed when describing the verandahs, 'between the actual bar bit an' then another bar ... if it wis a wean the could slip through, nae bother' (T2 L289). Likewise, one of the mothers told us 'the stairs are quite narrow ... maist o' them are a' broke anyway. Yer trippin up them at the best of times. Ah think we' a toddler ye'd be fear't tae let them go up and doon them' (G1 I1 L430).

In discussions about factors causing accidents, the following fixtures and fittings were identified as hazardous. The windows are seen as problematic for a number of reasons, the most pressing of which relates to the possibility of young children falling out of them. The risks are obvious, one teenager told us:

> When they're wee climbin' up tae see oot the windae or see whit's happenin'. They're always pullin' at handles an' a' that. It could happen quite easy. The windaes are'nae exactly hard tae open. Ye jist pull them (T2 L151).

Many homes have power points which are permanently live: not only do mothers complain that the fittings are loose and dangerous, but they note, too, that they are 'jist a perfect height for children tae put their fingers in' (G1 I1 L.645). The problem was graphically described by one of the teenagers

> ma wee brother usually tries tae pull the plugs when he's in his babywalker. He's just the right height tae reach the plugs ... 'cause there's no off an' on switch. It's jist live a' the time (T1 L770).

Other problems included fires, supplied without fire guards; metal sinks with sharp corners, cupboards without locks; and the positioning of the kitchen which often made it hard to watch children without exposing them to the risks of scalds and burns. The question of design came up several times: 'they're no designed with safety in mind at all' one mother told us 'an' yet they tell you that they're family hooses' (G1 I1 L718). Another suggested that part of the problem was a failure of imagination on the part of the architects.

> But maist o' these architects that build these houses are men – young men in their 20s and that and the day they're qualified, they go intae design, but they don't realize – they don't have a clue aboot havin' a family and different things. There's no really many women architects at all. Everything is male orientated and they just don't think about things like that. It's only women that think about things like that.

The wider environment, outside the home, was also described as being full of risks to child safety. Serious concern was expressed about the state of the back courts and bin areas which often meant that 'The weans cannae get oot tae play – dug's dirt ... it's disgusting' (G1 I1 L953). One teenager described Corkerhill's outside environment as 'mainly broken glass an' that, dog's dirt – fallin' in that or standin' in it. There's an' awful lot o' stray dogs up here so if yer runnin aboot or yer in a hurry it's a case o' that's you' (T2 L954). Likewise, a number of hazards were associated with road safety: not simply the problem of fast cars and careless driving, but also the risks added by stationary traffic obscuring children's view of the road, and by the debris associated with road works or housing repairs. Parents referred to a number of accidents which could have been avoided if road works have been properly protected and managed. According to parents, children face risks from the nearby railway, the river Cart, the clientele of the pub named after it, and even the designated play area, which had fallen into disrepair, was seen as dangerous.

iv) Safety is a parental and community responsibility, and this responsibility is taken seriously. The implications of the catalogue of risks summarized above are twofold: first, parents do know about the risks their children face; secondly, they have valuable experience in managing these risks, which prevention policies could build on. This is an important point because, given the very real dangers to children in an area like Corkerhill – dangers they clearly would not be exposed to in a more affluent part of the city – our primary interest should surely focus not on why accidents occur, but on why more children do not get hurt. Some pointers emerged from the group interview transcripts. Essentially, the reason why more accidents do not occur is that parents do see safety as one of their responsibilities, and they have already developed routines and strategies to fulfil this responsibility. As one mother explained

> ... And in the bath you've tae put the bath mat – well that bath mat I have for them is no good .... I've tae put towels, towels and more towels for the likes o' Malcolm tae stand on it or take a shower (G1 I3 L530).

In one case which was described to us, a parent's attempt to make the house safer came in for criticism from the housing department:

> Remember Mrs M? She had the big bit of fencing, 'cos her one went over (the verandah). Mind he got up and walked away. So she put a big screen up, a big lump o' wire, and she got into trouble for putting it up. She got told off. The housing said it wis an eyesore (G1 I2 Ls131–49).

It is clear that (as one might expect from people living in a risky environment) ways of handling potential hazards are already built into everyday behaviours and routines. This extends not only to responsibilities for a child in the family, but also to other children in the area. 'I mean, I would take responsibility for any o' they kids here ... it's no the first time I've taken them off the road' (G1 I3 L910). In some cases special care is needed.

> We had to watch Frank ... y'know blind Frank down the road. You had tae run out and take him across the road because there ... was a gap like that [where an unguarded hole had been left by workmen in the road] (G1 I3 L775).

Examples like this were routinely described to us. One mother remembered.

> I was cutting across the back and I looked up and this wean – she was about two – and the windae was wide open, and she was dancing there up at the windae, and if you batter at the door, she'd maybe get a fright, and I was at the letter box going 'Hello', trying to dae it as quietly as I could (G1 I1 L393).

Another described seeing children climbing the scaffolding outside her window:

> I mean I didnae even want tae open the windae in case I gied them a fright or something and they fell off ... I mean I'm two up and they're right up at my windae ... (G1 I3 L893).

The response to the question of adequate parental surveillance demonstrated a clear acknowledgement of the competing demands on the time of parents which is not always evident in the professional literature.

> You dae – I mean I would take responsibility for any of they kids here, and it's no because they've been neglected, but it's just that – maybe [the mother's] hanging out washing or something and in a split second, the wee one's away through the back ... (G1 I3 L911).

(v) Parents can and do moderate accident risks, but their role is limited in an inherently unsafe setting. Although parents do have strategies to keep their children safe, maintaining safety is, in the end, a question of cost as well as behaviour. In discussing methods and means of prevention, observations like the following recurred: 'Things like that ye huv tae buy them yersel. Any safety things in the house' (G1 I1 L718). As Table 2 shows, the costs of safety are not insignificant for people with relatively low incomes and high fixed outlays for rents, clothes and food.

The consequence of the rising cost of safekeeping is that concern about safety can add to the other anxieties mothers often feel about their child care responsibilities:

> you get really depressed and think about the things you children should have. You're told a' the time – y'know, you get these millions o' books when you

Table 2 The price of safety

	£
Socket covers	1.75 (for six)
Fireguard	19.99
Corner protectors	4.99 (for four)
Cooker guard	13.99
Curly kettle lead	8.50
Safety film for glass	17.99
Window catch	3.50
Stair gate	24.99

Source: Mothercare catalogue (1990).

have a baby and all this equipment you should have, and it can end up [with you thinking] 'I'm no bringing my child up properly' ... and [it] jist adds to all the other stresses you've got.

Conclusion

Much current work on accident prevention is exhortatory and is based on the idea that children and their parents can be effectively educated to avoid accidents. This view was borne out in our interview with the professional group (which will be reported elsewhere), and it assumes that the transfer of skills required to keep children safe flows from professionals to the public. Our group interviews indicate, however, that parental behaviour may be more noted for its role in mediating the risks in a hazardous environment than for enhancing those risks by neglect or fatalism. Parents are aware of the dangers facing children in their community; they already take steps to minimize these dangers where they can, and this is the point at which prevention policies could helpfully connect with local expertise.

Advice given by safety organizations suggests that risks can be reduced by:

(a) making sure that the home and garden are as safe as possible;
(b) teaching children to be aware of dangers and how to cope with them;
(c) never leaving children on their own;
(d) buying toys and equipment that are safety-approved and checked.

This is doubtless correct, but it ignores the social context of the lives of many children. Children do have to be left on their own from time to time when parents have (as they always do) competing priorities. Yet, making children aware of the dangers or educating them on how to cope with a risky environment can only ever be a partial solution.

In addition to relatively simple environmental modifications, one of the most significant routes to safekeeping identified by the parents, was the routine availability of social support. When there are competing priorities on adult time, responsible people other than parents are needed to care for children – while mothers are hanging out washing three storeys down, or going for messages, or responding to some other urgent need. Without extra help it may just be possible for a one- or even two-adult family to keep children indoors under a watchful eye in winter. But in the summer 'if everything was normal, you'd go out with them, but you cannae go out with them all the time' (G1 I3 L306). As for the alternative safekeeping strategy:

> You try to keep kids in one o these tenements tae play. What happens? Your neighbours get upset and that's it started, and then they're shouting at the wee bits o' kids and they're only doing what kids do ...

For mothers, the risk of accidents to children is an everyday hazard, and one which on the whole is successfully (if not easily) negotiated. Our research indicates that safety behaviours are learned (through shared as well as individual experiences) but that behavioural strategies are limited by factors associated with the design, character and condition of living space, and with the costs which individuals are expected to bear to make these environments safe.

Given the way in which safety behaviours are learned and shared, it can be argued that the 'top down' model of safety promotion will be less effective than a 'bottom up' model which builds on the expertise communities already have. Given that risks are not so much a product of parental neglect as a problem of coping with other household priorities in essentially dangerous places, effective accident prevention might depend on building a collective responsibility for safety into the rights and entitlements associated with the welfare state.

Notes

1 Group three, based on experiences in the cottage-type housing, was a smaller group and we were able to complete the group interviews with them in a period of just two weeks.
2 The first group were given a floor plan of their kind of dwelling and asked to mark dangerous places and features on it. However, as this group proved far happier talking about such features than they did drawing them, the remaining groups were not given a floor plan.

3 G refers to the group number, I the interview number, L the line number of the transcription.
4 The tapes were transcribed as spoken. To turn the transcriptions into 'Queen's English' would be to lose a great deal of the flavour of the interviews, and would be the equivalent of saying 'we could put it better' (which we could not).

The Public Health Research Unit is supported by the Chief Scientist Office, Scottish Office, Home and Health Department and the Greater Glasgow Health Board. The opinions expressed in this chapter are not necessarily those of the Scottish Home and Health Department. We are grateful to Cathy Rice and Betty Campbell and Walter Morrison for organizing the meetings described in this paper, and to Rita Dobbs for transcribing the interview tapes.

References

Blondel, B. et al (1985), 'Moralité des enfants de 1 à 4 ans dans les pays de la communauté européene', Archives Français Pediatrique, **42**:645–9.
Brown, G. and Davidson, S. (1978), 'Social class, psychiatric disorder of mother, and accidents to children', *Lancet*, 18 February, pp 378–80.
Burgess, J., Limb, M. and Harrison, C.M. (1988), 'Exploring environmental values through the medium of small groups: 1. Theory and practice', *Environment and Planning D: Society and Space*, **20**: 309–26.
Hedges, A. (1985), 'Group interviewing', in Walker, R. (ed.), *Applied Qualitative Research*, Aldershot: Gower.
Mitchell, J.C. (1983), 'Case and situation analysis', *Sociological Review*, **31**: 187–211.
Pharoah, P.O.D. and Alberman, E.D. (1990), 'Annual statistical review', *Archives of Disease in Childhood*, **65**: 147–51.
Ranson, R. (1987), 'Home safety – the challenge to public health', Coventry: University of Warwick.
Sibert, J. (1975), 'Stress in families of children who have ingested poisons', *British Medical Journal*, 12 July, pp 87–9.
Sinnot, R. and Jackson, H. (1987), 'Development in house and home safety', Coventry: University of Warwick.
Smith, S.J. (1987), 'Fear of crime: Beyond a geography of deviance', *Progress in Human Geography*, **11**.
Strathclyde Regional Council (1987a), *Child Pedestrian Casualties in Areas of Priority Treatment*, Glasgow: SRC.
Strathclyde Regional Council (1987b), *Voluntary Population Survey*, Glasgow: SRC.
Wadsworth, J. (1983), 'Family type and accidents in pre-school children', *Journal of Epidemiology and Community Health*, **37**:10–14.

Index

For Product Safety Concerns and Information please contact our EU representative GPSR@taylorandfrancis.com
Taylor & Francis Verlag GmbH, Kaufingerstraße 24, 80331 München, Germany

www.ingramcontent.com/pod-product-compliance
Ingram Content Group UK Ltd.
Pitfield, Milton Keynes, MK11 3LW, UK
UKHW022156190926
475702UK00003B/55

* 9 7 8 0 8 1 5 3 4 8 6 9 6 *